"The book is written in an informal style, embellished with anecdotes; but it offers good practical advice about preliminary research, locating sites, conducting investigations, scientific record keeping, preservation of finds, publication of reports, and equipment which may be made inexpensively. Recommended." —*Library Journal*

"A serious book, loaded with useful information."
—*Times Literary Supplement*

"The author has had years of experience in British and Mediterranean waters. The subject of study includes wrecks, harbors, sunken ports, etc., which may be found in rivers, along beaches, and in the open sea — in fact, almost anyplace where there is water."
—*American Reference Book Annual*

"A handbook for skin divers containing information necessary for both large and small-scale archeological expeditions beneath the waters."
—*Charleston Evening Post*

A HANDBOOK OF UNDERWATER EXPLORATION

A HANDBOOK OF UNDERWATER EXPLORATION

Bill St. John Wilkes

STEIN AND DAY / Publishers / New York

FIRST STEIN AND DAY PAPERBACK EDITION, 1975

First published in 1971 under the title *Nautical Archaeology*
Copyright © 1971 by Bill St. John Wilkes
Library of Congress Catalog No. 77-159558
All rights reserved
Printed in the United States of America
Stein and Day/*Publishers*/Scarborough House,
Briarcliff Manor, New York 10510
ISBN 0-8128-1807-5

CONTENTS

	List of Illustrations	5
	Preface	7
1	Archaeology in the Diver's World	11
2	The where, why, and how of underwater archaeology	18
3	Preliminary research	34
4	Preparation	50
5	More advanced equipment to make	78
6	Search techniques	95
7	The signs we seek	129
8	Locating and pinpointing the site	140
9	Recording underwater finds	158
10	The place of photography in surveying	180
11	Underwater excavation	202
12	Dating and identification	238
13	Conservation and reporting	248
	Bibliography	270
	Appendices	277
	Index	291

ILLUSTRATIONS
Line drawings by the author

Plates

The 1969 Mensura Diving Team expedition to Bithia	65
Setting up a base line theodolite station	65
Working in the quicksand surrounding the *Amsterdam*	66
A high level view of the *Amsterdam*	83
A slotted angle grid	84
Suction dredging	84
A heavy scaffolding pipe grid (*George F. Bass*)	181
A high pressure water jet	181
Foundation blocks on the site of Bithia	182
A collection of amphoras	199
The quay/quarry area of the Roman port at Tharros	199
Bronze cannon raised from the *Amsterdam*	200
Divers taking bearings and measurements	200

In text

Drawing board; detail of paper clamp; 50m survey base line	54
Floating marker; bottom marker	57
Survey floats; counter weight system	60
Underwater search grid	63
Underwater scaffold pole grid	68
Diagrammatic sketch of the 'Morrison' grid	70
Underwater measuring tools	74

Setting out a secondary base line; optical square Hilger & Watts Ltd	76
Two versions of plane tables	79
Combined theodolite and U/W survey table	86
Underwater levelling device	93
Swimsearch techniques	97
Two suggested methods of carrying out a sweep search using divers	101
Devices for towing divers	104
Details of a simple underwater sled	106
Search patterns for close cover	110
Coring equipment	116
Typical echo sounder indications	124
Illustrating use of transits	142
Compass bearings from site to land; bearings drawn on chart	144
The main features of a sextant; using the sextant to fix a position	146
Establishing survey stations and base line; tacheometry	150
Triangulating a site	163
Form showing finds	165
Taking levels underwater	169
Mosaic photography; camera stand and grid	185
Use of 1m square or cross for survey work	191
Photogrammetry equipment; detail of cross; inclinometer	194
Projecting a grid from a tilted square	196
Pressure jet nozzle; harnessing currents to excavate	207
A basically simple airlift	212
Underwater dredge	222
Lifting equipment	228
Use of buoyancy bags	234
Casting a reproduction; encapsulating a small object	253
Drawing anchors	264
Drawing a classical find	266
Drawing cannon	268

PREFACE

MOST DIVERS will probably feel that the possibilities of a small group of divers being able to undertake worthwhile expedition work is very remote. The underwater films we see on television may well arouse our interest but it is no doubt tempered with the realisation that tremendous resources are behind those very wonderfully equipped teams. For the majority of us, such enterprises are well beyond our reach, so this book is aimed at helping those same small groups to achieve results every bit as interesting and every bit as valuable as those shown on television, but at a minute proportion of their costs.

I would also like to offer this book as an indication of my appreciation to all those who, by word and deed, have helped my own expeditions, upon which this work is based. In particular I would like to dedicate the book to the three people who have been paramount in making the whole thing possible —to Dr Nic Flemming who originally suggested the idea of doing these expeditions and who has been a source of great assistance behind the scenes, to Miss Joan du Plat Taylor, who has been a constant source of encouragement and assistance— and to Ann, my wife, who has borne the brunt of the work ranging from co-driving and co-diving, through marathon feats of camp cooking to having to type the final manuscript.

Bill St John Wilkes

A HANDBOOK OF UNDERWATER EXPLORATION

1
ARCHAEOLOGY IN THE DIVER'S WORLD

OVER THE last decade underwater sport has increased from something enjoyed by a mere handful of divers to a major recreation which is being catered for by a wide range of organisations in Europe and America. By the setting of rigorous standards of training and proficiency, most of the hazards of the sport have been minimised, but it must be accepted that whenever man steps out of his natural element, there must be risk. Only by insistence that the novice diver should progress gradually through his or her training, stage by stage and under strict surveillance until he reaches a satisfactory standard, can this risk be reduced to the acceptable.

The need for training is indisputable. Practically anyone can don mask, aqualung, and fins and swim under water, it is not even necessary to be a good general swimmer. Provided a person is reasonably fit, has no ear problems, and is not claustrophobic, he can probably get by under water. But this is not enough, a diver must be prepared for the unexpected and that is where the training he has absorbed becomes invaluable and even possibly lifesaving.

There is a feeling amongst some club divers that their diving officers and trainers try to make it too difficult. One gains the idea that the passing of tests is on a par with the Red Indian ritual initiation tests which were carried out by a young Indian before he was accepted as a brave! Make the tests as difficult as possible, even frightening to the uninitiated and only the fittest, bravest and best will pass. True thinking, to a degree, but awfully wasteful, for many trainee divers lose their enthusiasm and give in before they have learnt that there is much in store for those who do become proficient.

This attitude is by no means general. By far the majority of trainers are doing a superb job and were it not for their enthusiasm and dedication we should have far fewer divers. Part of the problem is due to the fact that so much training must be carried out in swimming pools, many of which are so antiquated that they are barely hygienic and offer totally inadequate facilities. The multiplication of swimming pools over the past five years has been encouraging and we are now able to use pools of Olympic standard, but excellent as these are, they cannot duplicate the one essential ingredient which the open water of the sea or lake can offer—interest.

The poor diving officer is put to his wits' end in trying to stimulate enthusiasm, but even the best of his 'party tricks' begin to pall after a time. While there is the challenge of tests to be passed, many will persevere. But after this, with open-water dives limited by lack of accessibility and inclement weather, it is little wonder that enthusiasm flags and only the more stoical divers carry on. After all only so many divers can train more divers to train more divers to train . . .

The essence of really successful pool training and club management is to accept personally and to then promulgate to others the idea that diving is only a tool, a tool to be used with skill in carrying out underwater assignments. It is the transport to get us to the site of our task, just as in a car or a motorcycle.

Even with the carrying out of the initial training, attempts should be made to introduce tasks, games, or tests that will lead to useful work under water at a later date. Once a diver has achieved his certificate of proficiency he immediately needs the stimulus of other activities to use as an extension of his diving. Clubs should be constantly on the lookout for projects which will provide interest, experience, and even possibly revenue. Many clubs augment their income from subscriptions by helping fishermen and yachtsmen recover gear and in so doing help to bridge the gap, which still exists in some areas, between the groups.

Underwater photography alone is an enthralling pastime to

Archaeology in the Diver's World

which many divers are already conditioned by their above-water interests. It is a subject that requires complete concentration under water, a great deal of knowledge, and lots of practice. Much of this can be taught in a pool and many clubs do so. But there are the limitations of expensive equipment and poor open-water visibility which between them will restrict to some degree, the feasibility of this pastime.

Other clubs team up with geologists and biologists, producing for them valuable results they could not otherwise hope to obtain. A very worthwhile activity and one that is only limited by time and energy plus the desire to seek out and approach bodies that can utilise their services.

There is also the 'sport' of underwater spearfishing, but this does not or should not involve aqualungs and is relished by a comparative few. Skill and endurance are certainly prerequisites here; nobody can deny it; but this is the sport of the hunter and not of the diver. It is one thing to catch fish for food but another merely to catch for the sake of catching to get the greatest take.

These and many others are post-training activities that clubs can participate in, but in addition there is one other which must stand supreme—underwater archaeology. For to carry this out, all other skills are needed. The skill of the hunter, and his patience, in searching for clues; the knowledge of geology, if only to distinguish the man-made structure from the natural indigenous rock. Many times this will be the first clue that one perceives and a skilled eye is required to do it.

Photographic and drawing abilities, essential to record finds, and general seamanship, to study the effects of tides, currents, and navigation, are all involved in archaeology. Expertise in archaeology itself is not essential in preliminary work although preparatory reading, visits to land digs, and research will always pay handsome dividends. Apart from exercising so many interests, underwater archaeology has another major advantage—it can be carried out practically anywhere where there is water! In the sea, in lakes, rivers, and even down wells: there is no limit.

The coast of every traditionally seafaring nation boasts an almost uncountable collection of wrecks. The examination of these can fill the quite remarkable gaps in our knowledge of the building and equipping of ships built within quite recent times. Even as recently as 100 years ago, ship-building was an art rather than a science. Skills were passed from father to son, drawings were incidental and were mainly artists' impressions. Specifications merely detailed size, weight and sizes of main timbers—the rest was done by thumb and eye! The finding of substantial portions of ships can tell us much.

Of course, the mere word 'wreck' calls forth visions of treasure and certainly magnificent hauls have been made. The temptation to go hell-for-leather in salvaging such finds can be almost irresistible and can easily overcome any desire to excavate the wreck more systematically, so that the archaeological and historical values can be learned. The use of explosives on wrecks is an example of this. Capably and knowledgeably used, explosives can be an asset, but used indiscriminately, they can be devastating in their effect and much of value can be destroyed.

In 1964, the Committee for Nautical Archaeology was formed in London with the object of co-ordinating divers and land archaeologists, for so long suspicious of each other! Since then great progress has been made in this direction: competent and responsible divers are accepted as being capable of contributing to archaeology. Many archaeologists now welcome and indeed seek the support of recognised teams. Additionally, the committee is investigating methods of protecting underwater finds from unauthorised excavation (plundering). By this it is not meant that the CNA are trying to impose a general restriction on divers, on the contrary, it is their aim to support divers who do locate finds of value and to provide assistance, both technical and financial, to enable them to carry out their work. At the time of writing there is no comparable body overseas, but it is hoped that the example set in the UK will be followed by others.

In any event, each and every wreck is considered to belong to some party. Under the laws of salvage, which we shall discuss later, this is clearly established and it is the responsibility of the diver, and very much in his interests, to endeavour to contact the owners and make his arrangement with them.

The *de Liefde*, wrecked off Scotland, was an East Indiaman and the ownership was passed to the Dutch Government upon the bankruptcy of the Dutch East India Company in the late eighteenth century. The *Santa Maria de la Rosa*, wrecked in the Blasket Sound, Ireland, is still the property of the Spanish Government and rights to salvage were obtained by Mr Sydney Wignall, who successfuly located the wreck. A 'pirate' venture tried to get in on the salvage but a court injunction was issued against them and prevented them from so doing. The *Association* wrecked off the Scilly Isles, was an Admiralty vessel and therefore salvage rights had to be obtained from the Ministry of Defence (Navy). Unfortunately, this was prior to the CNA having knowledge of this wreck and the MOD awarded salvage rights to three separate groups at the same time and without any limit of time. The result was not conducive to an archaeological excavation being carried out! As a result, the MOD will not now issue salvage rights to any group for work on wrecks likely to have historical or archaeological value without prior reference to the CNA, who will advise them of the suitability of the groups requesting salvage rights to carry out the work in an archaeologically satisfactory manner.

Apart from the glamorous subject of wrecks, archaeology covers many subjects, for instance the sunken ports in the Mediterranean or on the British coastline, such as at Fishbourne in Sussex. Major Hume Wallace's work off Selsey has found numbers of catapult or ballista missiles on the sea bed, which are indicative of a fortress possibly positioned there to protect the harbour to that town. Major Wallace also suggests that it may equally have been a quarry from which these missiles may have been hewn. There is a remarkable similarity between those found in the sea at Selsey and those found around the Roman castle at Pevensey, some fifty miles away, indicating a link between them.

In the Scilly Isles and in some Scottish lochs we have evidence of ancient post villages (Crannogs) and the remains of wooden posts or piles can still be found under the water. Work has been done on these but much more remains to be carried out. Rivers are a constant source of finds of value, both intrinsically and historically. The burnt out hull of the *Grace Dieu*, now more of a land excavation than a dive, awaits the attention of archaeologists, the sites of the Roman bridges across the Thames are being sought and the ancient shipyard at Bucklers Hard at Beaulieu is yet another instance.

In the Western hemisphere, Port Royal, Jamaica is an outstanding example of a sunken city, again due to earthquake on 7 June 1692. Virtually the whole Atlantic seaboard of the US is rich with wrecks of all nationalities and, inland, the Great Lakes and Lake Champlain all have their share of interesting wrecks.

So the list goes on—and so opportunities for useful, enjoyable, challenging, and satisfying underwater projects can be found by those willing to make the effort and not just to stand around carping at the lack of worthwhile projects.

But remember, underwater archaeology, like land archaeology, is very similar to police detection—99 per cent perspiration and 1 per cent inspiration. Before any real knowledge of archaeology is required, a tremendous amount of really hard, frustrating searching must take place. This only requires imagination, determination, patience, and fitness. This latter point must not be overlooked. The physical and mental effort of searching under water is far greater than a casual dive. Actual work in detail on a site is an even greater stress and complete physical fitness is essential.

Once the initial spadework has been done, it is then very important to have a knowledge of archaeology or at least of archaeological methods and techniques. In Britain the CNA will willingly help here and with sufficient prior notice will generally be able to provide a speaker at meetings. Also under the sponsorship of the CNA a series of week-end and one- and two-week courses on underwater archaeological techniques is available at the School for Nautical Archaeology at Plymouth (SNAP) under the guidance of Lt-Cmdr Alan Bax and Mr Jim Gill. This school is quite unique, but again it is hoped that in time other countries will offer similar facilities to diver-archaeologists.

Having read of expeditions by Cousteau, George Bass, Peter Throckmorton, and others, many will think that the requirements of equipment, time, and resources make it impossible for the amateur group to undertake similar work. Well, of course this is so—at least as regards extent—but much work can be done 'on a shoe string' and this book is setting out to help you to do so. Limitations of time, equipment, and money

Archaeology in the Diver's World

can be very largely overcome by improvisation, ingenuity, and most important, by planning.

The following chapters may help you. It is not suggested that the ideas set out are the only ways of accomplishing a given task, or even the best—but they do work!

2
THE WHERE, WHY, AND HOW OF UNDERWATER ARCHAEOLOGY

WHEN DOES a find have an archaeological value? Is a find of value archaeologically simply because it is old? If not, how old has it to be?

The general dictionary definition of archaeology is the 'study of antiquities or prehistoric remains' which does not help us at all, for this indicates remains of many centuries ago, of earlier civilisations and cults. But today, even in land archaeology, thought is being given to the much more recent past, even to the extent of the recovery of machinery, etc, of the Industrial Revolution. Inevitably, as every year passes, the modern pace of technological advancement makes yesterday's miracles obsolescent and those of ten years ago—obsolete. So, generalising, perhaps we can say that any find that throws light on ways of life, techniques in construction or shipbuilding, methods of travel and travel routes, can have an archaeological value, irrespective of how far back these date.

Taking another logical step, we can say the recovery of objects that were known by report or even drawings but not previously found intact, could also fill the bill. Referring to the Industrial Revolution again, much is known about this period from contemporary writings and drawings, but the wholesale destruction of buildings and machinery has erased nearly all physical remains of the time. Hence the emphasis now upon treating this period as an important archaeological era.

For our practical purposes, let us assume the most recent time of our 'archaeological period' to be about 100 years ago.

The where, why, and how of underwater archaeology

Certainly, when talking of ships, this becomes quite an important time. For this was the era of transition from sail to steam, from wooden hulls to steel and, more to the point, it was the time when the art of shipbuilding mentioned in the last chapter began its change to the 'science of shipbuilding'.

From about this period, far more detailed drawings of actual construction were used and have been retained. Consequently naval historians have a fair degree of detailed knowledge of the shipwrights' methods after that time, whereas from the earlier times only rudimentary details can be gleaned from contemporary writings and from artists' drawings.

Many readers will have visited Nelson's flagship, HMS *Victory*, in dry dock at Portsmouth and will have been escorted on guided tours over her, marvelling at the small dimensions of the ship as a whole and listening in awe to the guides' colourful description of the working and living areas. The difference in comfort between Nelson's own quarters and those of the men is only marginal, and serves to impress the viewer with the hardiness of the men who sailed the ship. But from our point of view as archaeologists, how accurate is the statement that his quarters are 'just as they were in his time' with all the accoutrements exactly the same. This is a very moot point. For it is certain that within the ship as a whole, there have been many changes made from the day she was commissioned to the eventual day when she was laid up in her present dry dock. In today's modern navies, complete refits have been made to many naval units, which, even if they have not altered their character entirely, have certainly altered the internal construction and, even more certainly, have altered the fittings and equipment. So it is very probable that, looking at *Victory* today we see a ship vitally different, at least in detail, to the other ships of her day, of which we have no further examples.

The long search in Stockholm harbour, leading to the finding of the *Wasa*, (not *Vasa*, as it is often spelt) and the eventual skilful and costly recovery has paid enormous dividends in increasing our knowledge of shipbuilding of her time. This ship was lost on her maiden voyage and therefore went down exactly as she was constructed; within her hull were all of her fighting equipment and fittings as well as the crew's own personal possessions. On her recovery, all these details were found, even the foodstuffs with which she had been victualled.

Naval historians and architects have, in that one remarkable find, obtained chapter and verse, the most minute details of shipbuilding, ship fittings, and ship's life on a capital ship of her period. This information could have been obtained in no other way. The story of her raising is well told by Commander Bengt Ohrelius in *Vasa, The King's Ship*.

On English shores, of the multitudinous tales of wreck that have survived over the centuries, few, surely, can be stranger than that of the *Amsterdam*. Her story commences in the almost traditional way, which usually tells of hardships at sea, the constant unavailing fight against tempest, sickness, death, and even perhaps mutiny, before the violent tragic end on some hard, forbidding rocky and unyielding coastline. But there is a difference: her end was on a very yielding coast!

So let us cast back to the year of 1748, for it was at the end of that year that a fine, new, 54 gun Dutch East Indiaman set sail for Batavia in the Dutch East Indies. A proud ship, one of the largest of the fleet, she carried a cargo of all manner of goods needed to cater for the needs of an expanding colony. With a tonnage of 700, length of 150ft, beam 35ft, her complement consisted of 333 sailors and soldiers, officers, and three women passengers.

Three times she set sail, twice she had to return for repair and alteration in the face of atrocious weather. Eventually, about Christmas time, she finally left Texel at the mouth of the Waddenzee where she had been sheltering, for her third and last attempt. Eighteen days later, constantly tacking and beating against raging winds, she still had not passed Beachy Head on the Sussex coast. Some fifty of her crew had been lost through falling overboard and from illness. Some say that the yellow fever had been brought aboard by a crew member signing on after a voyage in the tropics. Eventually, in despair, her commander, Captain Williams Klump, turned her bows and ran for the shelter of the English coastline.

Possibly hoping to anchor in the lee of Langney Point, which juts out and protects Pevensey Bay, the *Amsterdam* struck and lost her rudder. Now, out of control and at the mercy of the storms, she swept back along the coast, past Normans Bay and

Bexhill until, at last, off Bulverhythe her anchors held her fast and her crew could rest awhile. This would have been about 10–11 January 1749. She was there for several days waiting out the storm and trying to rig a jury rudder to enable her to go on to Portsmouth for proper repairs or possibly even to return to Texel again. However, fate was against her and at 3 pm on a Sunday afternoon she drove on to the shore.

An interesting series of letters addressed to one John Collier of Hastings relates this and subsequent events. Apparently as she approached the shore there was the sound of many guns firing and drunken singing! The latter is not surprising as her manifests show that she carried 'many thousand dozen bottles of wine'. Whether she was deliberately run ashore we cannot say for certain, but it could be that a drunken crew decided to beach her and mutinied. Spent musket balls which had flattened themselves against her planking may have come from the guns that were heard to fire. These were just some of the interesting articles which were to be found on her.

Whatever the cause, the *Amsterdam* ran bows-on on to the beach and struck soft sand some 200m below high-water mark. As the tide fell the rocking of the hull in the storm which still raged and her great weight caused her to sink rapidly down into the quicksand. A most yielding coast!

On 17 January, George Worge wrote ... 'she stands in a good place, and in appearance quite whole, and may do so for some months ...' But on the twenty-fourth Richard Patrick wrote '... the Dutch ship I mentioned to you still sits whole, and the plunderers speed but Indifferently, neither do the Owners save any Quantity of Goods, for the Ship is so much Swerved in the Sand, that it is Impossible to get at the Cargoe, the Ship being always full of water ...' So, in one week, she had settled so that at high water all decks were awash and even at low tide, her lower decks were under water.

Despite salvage attempts then and later, even into the nineteenth century, it seems likely that her main holds were never entered. But it is recorded that her money chests, numbering twenty-eight in all, containing silver coin and bullion worth £60,000, were taken ashore and safely carried to London. But one chest turned out to be empty—somebody had got there first!

Drastic means were resorted to in the attempt to get into the holds, including building enormous bonfires on the deck—which were always put out by the waves! Even gunpowder was used in an attempt to blow up the decks, but again with little results except that a Mr Nutt, engineer, apparently did not retire quickly or far enough after lighting the fuse and so blew himself up, poor chap.

So the proud *Amsterdam* quietly settled into her bed of quicksand and clay, her timbers nestling amongst the peat beds and fossilised trunks of the ancient forest that once stood over this area. At very low tides, some of these tree trunks and between them the ribs of the Indiaman could be seen still. But she was forgotten and just lay there waiting for someone to come along to ask her to give up her secrets.

Ken Young was the 'someone'. He was also the site agent for Wm Press Ltd, contractors, who were laying the three-mile outfall for Hastings new sewage disposal close by. Ken was also interested in history and wrecks and had made a practice of researching any area in which he was working. He had come across the 'Collier Letters' and decided to look for the *Amsterdam,* and soon found her only 180m from the outfall. Only the tips of broken ribs showed above the sand, sketching the outline of her shapely hull on the smooth sand.

Curiosity is a powerful thing and sometimes needs powerful tools to satisfy it. With only some forty minutes of exposure at low water, speed was all-important and Ken had just the tool—a mechanical digger. Not the ideal archaeological tool, but who cared, 'let's get into the hull and see what was there', was the feeling at the time, for these men were not archaeologists.

Sitting alongside the ribs, the powerful machine was soon gulping great bites of sand, slurry and water from deep inside her. One, two, three scoops produced no results, but the next hit the jackpot! Up came bronze cannon, still wrapped in the muslin which had protected them as cargo. Then blocks and tackle, one with a still free-running sheave—then a vast range of items, wine bottles, gin jugs, the last for a child's shoe, cutlery, plate, smoothing iron, flute, ladies' combs, fans, and an enormous number of clay pipes, some with tobacco still in the bowls. And so the list went on, four foolscap sheets of items listed line by line. Sadly, some ninety per cent was broken, through the use of the mechanical digger. In fairness, it should be said that this would have never been used had it been thought for a moment that there might have been such items inside the ship.

The importance of the finds now struck home, the curator of Hastings Museum, John Mainwaring Baines, was called in, the finds were reported to the Receiver of Wrecks, and conservation commenced. Next a meeting of the representatives of Wm Press Ltd, their solicitors, and representatives of the Committee for Nautical Archaeology was convened to discuss the best way to handle the situation.

Obviously, no further work could be indiscriminately carried out using heavy equipment and without it excavation is impossible, for the sand fills holes as quickly as one can dig them. It was therefore decided to carry out a thorough examination of the hull on the next equinoctial low tides.

It was evident from the finds already made that we were working upon a wreck of more than passing importance. This was confirmed by Professor Van der Hiede from Holland, who came over specially to see the hull and the finds. He has stated that he believes that here is a wreck of equal importance to the *Wasa* of Sweden. As this was a distinct likelihood, it was necessary to establish the condition of the hull's timbers, so that plans might be made to lift the ship and her cargo. Timber pulled up from beneath the sand on the earlier 'grab' proved to be as sound as new. If this were to be the same throughout the wreck, it would be wonderful.

So plans were made to dig down close to the walls of the ship, both inside and outside, to examine her planking. Because we had so little time on each tide we reluctantly decided to use the mechanical digger again. Here tribute must be made to the operator, for he handled the cumbersome machine so carefully that he was able to scrape the sand off the sides of the hull to within an inch without scratching the timber. Even here we had problems for the sand was so soft in places that the machines sank down over their own tracks in a matter of moments. It was impossible to walk in some places, let alone stand still, for one quickly and frighteningly sank deeply into the quicksand.

A surface plan was drawn by Peter Marsden and from this it became evident that the hull had a list of approximately 15 degrees to port. This meant that although equal amounts of ribs were standing proud of the sand on each side, those on the starboard side were, in fact, lower down the ship than those on the port side. The starboard timbers had been rotted away down to approximately midway between the two gun decks, while the port side was level with the first gun deck. This helped us in a way, for by digging down on opposite sides of the hull and taking the hull profile, that from the port side reflected the hull section at a higher level than that on the starboard. The two could be matched and gave an overall profile equal to approximately 16ft down, a depth to which it would have been impossible to dig.

Even with the mechanical digger and a 6in pump going full blast, it was no easy task to keep the holes from filling in and, indeed, to prevent the heavy equipment from itself falling in. Full dry suits were worn when working in the holes for, apart from being very, very wet it was also more than slightly pungent—a still active sewer outfall discharged some hundred metres away up-current from the site!

On the port side we got down no more than 2.5m and in the process uncovered two gun ports with their hatches still closed, several scuppers, a chainplate that once held the rigging and still retained pieces of rope, and also a large collection of clay pipes and handfuls of tobacco. The presence of so many pipes was interesting and may provide support for the belief that sailors of old smoked tobacco and herbs when disease was rampant on their ships. A very interesting 'scarfe' joint was found as well, in two of her planks.

The timbers were found to be absolutely hard and sound, but their fastenings had, apparently, been rusted away and the planks were inclined to move. At one time work was proceeding at the bottom of the hole when a plank suddenly 'started' and water gushed out from inside the hull and quickly filled the hole—uncomfortably quickly, it may be added.

On the starboard side it was possible to dig deeper, to about four metres, for at just over one metre down a very hard bank of dense clay was encountered and digging in this was easy for the sides did not keep collapsing inwards. It may well have

The where, why, and how of underwater archaeology

been this bank of clay that caused the ship to heel over as she sank in the otherwise soft sand. The planking here was also in excellent condition, but again the weight of sand and water on the inside caused the whole of the area to move outwards some eight to ten centimetres. Only quick action in removing the sand inside the hull by means of the digger and so reducing the loading on the wall, prevented possible extensive damage. Later, when the tide had turned, it was noticed that the hole within the hull filled with water more quickly than the hole on the outside which had been dammed. The pressure of the water on the inside forced another plank to move and it sprang outwards several centimetres.

Amidships, we dug down about 2.5m without finding anything of note. Two vertical posts appeared, which were joined by horizontal planking, and this may have been part of a hatchway. We could feel solid timber about half a metre below the lowest depth to which we dug, but could not get down to it. This may well have been the lower gun deck.

Working very hard through some nine low tides, sufficient information was obtained to satisfy us that here indeed was a wreck of great archaeological importance. For locked within the holds of the *Amsterdam* is as great a cross-section of household goods, personal apparel and equipment, shipboard goods and accoutrements of the 1740s as is ever likely to come to light. At the same time the hull itself seems as complete as that of the *Wasa*, and is the only known remaining hull of an East Indiaman.

It is probable that, to excavate this vessel, it will be necessary virtually to dry dock her within a cofferdam and then carry out a dry-land type of dig, layer by layer, replacing the plank fastenings as each plank becomes clear. For it seems that it is only the sand within and without the hull that is keeping it in shape; remove this and one might have an interesting pile of timbers but no hull!

To turn to an older example in England: in the bed of the Solent lies the ill-fated *Mary Rose*. This fine ship, which was to have been the most powerful ship of the British navy, was one of the first to be equipped with batteries of heavy guns. Her design embodied all the most up-to-date features of her time. On her maiden voyage as a fighting unit, she sailed out to fight the French in 1545. As she passed Spithead before the king and queen, her complement of crew and soldiers lined her bulwarks to salute, a sudden gusty squall hit the high topsides of the vessel, and this, coupled with the imbalance of all her personnel on the one side, caused her to capsize with a tremendous and tragic loss of life. Italian divers were later brought over and removed the upper works of masts and rigging which threatened to foul other ships using the Solent, and the hull was left to sink into the bottom.

In the nineteenth century Royal Naval divers carrying out practice underwater demolition work on another wreck, that of the *Royal George*, sunk close by, were on one occasion swept by the tide down on to the *Mary Rose*. But since then, she has defied efforts to find her. A *Mary Rose* (1967) Committee has been formed, supported by the CNA, and working with divers from the Southsea Branch of the BS–AC, under the guidance of Alexander McKee and Mrs Margaret Rule of the West Sussex Archaeological Society, are carrying on this work which has been fully detailed by Mr McKee in his book *History Under the Sea*.

There are also the galleons of Spain, including the wrecks of her Armada on the UK coast, and the many wrecks of the heavily laden convoys returning from the New World. Many of these ships carried valuable treasure in coin, jewellery, and plate and are targets for treasure hunters in both the Eastern and Western hemispheres. Few of us can resist the feeling of excitement when we read of the discovery of a new hoard found off the Keys of Florida, the coast of Ireland, or in the Scillies. Many a fortune has been lost in trying to locate such treasures; more, probably, than has really been found to justify it. Unfortunately, many lives have also been lost. Certainly there is no harm in seeking bullion, etc, but let us always try to remember that the actual hulls of these sunken craft and the odd artifacts of equipment found on them are the things which really help tell a story and complete a few more details of the jigsaw portraying our knowledge of the ships of all eras.

Further back in time still, we come to the wrecks of the Romans and the Greeks. Anyone interested in archaeology has heard of the amphora, the ubiquitous 'carry-all' of those

The where, why, and how of underwater archaeology 27

ancient peoples. These exotically shaped containers were really very functional: they carried all loose and valuable commodities including wines, oils, water, cereals, seeds, valuable minerals, and even money and jewellery. They were the ancient equivalent of the plastic container of today, the tin can of yesterday, and the barrel before that. To the archaeologist, their value does not stop there.

Amphoras were always locally made to suit the needs of the local people. They were therefore made from local clays, in patterns peculiar to the locality and even the detailed design might differ at the whim of a particular potter. Consequently, the point of origin can be readily established by comparison of shape and material. The technique of making them was passed from father to son but differences of design did occur, influenced by changing trends of the different periods. It has therefore been possible to catalogue large numbers of designs (Dressel and others—see bibliography) against which a comparison of a new find may be made. From this a fairly accurate dating is possible. So the finding of one amphora type in a wreck can at once tell us from whence the amphoras, and therefore logically, the wreck, orginally came and the period at which she sank. Additionally, it was quite common practice for the potter to put his own mark on the amphoras, generally on the shoulder or the rim, and this establishes a precise locality of origin. Finally, the shipper frequently marked with his own seal the clay or pitch stoppers used to seal the containers. This was presumably as a precaution against the skipper of the craft taking a few hearty swigs of the contents and replacing them with water! Whatever the reason, the seals provide one more piece of information for identification.

As more wrecks are found, each bearing cargoes of amphoras, charting of these can trace out trading routes of various shippers, and show how these ancient mariners gradually extended their net of trading over what were for those times, vast areas.

The search for the wrecks for their content or cargo value alone is archaeologically rewarding, but the hulls themselves are still a closed book as far as construction is concerned, certainly in detail. For instance, it is known that the Romans built their ships with planks in a carvel form, a horizontal layering of planks edge to edge. We still do this today, but an interesting difference appears to be that they used short wooden dowels, ie little round wooden rods let into drilled holes along the edges of the planks to hold them more firmly together. This is a feasible method of construction when the planks are first laid up, but what happened if a plank was replaced? It seems unlikely that the dowels would be replaced for this would mean removing the adjacent planks all the way up the side of the craft. If they did not bother to replace them, why use dowels in the first place? The finding of a hull with a portion of the planking bearing signs of replacement might shed some light on this point.

Going further back we come to Bronze Age ships like that found at Gelidonya which has been so well reported by George Bass and Peter Throckmorton in their book *Excavating a Bronze Age Shipwreck*.

One day, we hope, it may be possible to date a wreck very accurately by the annual growth rings shown in the large timbers used in the ship's construction. It has already been established that the growth rings of a tree vary according to the type of year in which the growth occurred. In a warm, moist season the growth will be greater than in a dry, cold one, consequently the width of the ring will be that much greater in a warm year. It appears that these ring variations are very consistent over a particular climatic zone and all trees within that area will have the same pattern or sequence of thick and thin rings.

It is hoped that it will be possible to form a complete chart of ring 'thicknesses' going back over very many years. This is being done by charting ring widths of baulks of timber known to have been cut at a particular time. With this date known, the number of rings indicates the actual length of time the tree had been growing and so a span of years may be recorded in the one timber, indicating the variations of good and bad seasons over many years. The ring widths or thicknesses are recorded as a 'saw-tooth' graph. By making a similar graph of a sample piece of wood that has been found and comparing it with the master graph, similarities of pattern will be noted. Provided these span a sufficient time interval one can say that their growth period was at the time indicated by the known dates on the master graph.

The where, why, and how of underwater archaeology

For instance, a baulk of timber from a wreck is found, at the same time coins are also discovered. In certain circumstances (but see p 244) it may be assumed that the wreck occurred at a date soon after the latest dated coin. The ring pattern of the timber can then be said to have antedated the coins by some years, ie, the age of the craft, plus the life span of the tree in growth. So immediately one has a pattern of growth ring thicknesses for a known span of years. This can be converted to a graph and offered up against a master graph. If the patterns match at some point, the year of commencement of growth and the year of cutting of the timber can be established. On the other hand the overlap may be small, thus offering the means of extending the graph still further.

By collating numerous patterns, a master graph covering many hundreds of years can gradually be completed, against which future finds may be matched and dated. As trees from different climatic zones can have different growth patterns, it is essential to ensure that all samples used for a particular graph are taken from the same zone. This technique is still in its formative stage and is being developed in America, but obviously, it will take many, many years before a span of even one thousand years can be completed.

So far we have only talked of wrecks, but of equal importance to the archaeologists are the remains of other works of man. These can range from subjects as vast as complete towns and harbours to diving down wells! Such subjects do not have the same glamour as that associated with wrecks filled with loot, but are of considerable importance. To the diver, they offer a similar satisfaction in seeing something which no man may have set eyes on for perhaps 2,000 years. They may not, however, hold the dubious thrill of diving to any great depth, as is frequently the case where wrecks are concerned. One so often hears the remark, claim, or boast that 'we were down 100ft' as though this in itself was a remarkable achievement. The writer's own experience has convinced him that water is just as wet at 30ft as it is at 100ft! It is also usually much more enjoyable to dive to the former depth with warmth, visibility, and endurance of dive enhanced proportionately.

In 1968 the divers with the author's team logged over 900 hours between ten divers over eighteen days. This could never have been achieved if we had been working at any great depth, but where we were, good, interesting dives were enjoyed, worthwhile work was completed and photography, etc was carried out in natural light; and decompression was never a concern!

In the Mediterranean there is tremendous scope for this work: there are nearly 280 historically recorded sites on the various seaboards of the Mediterranean. Some of these will never be located, some have been built over by modern cities but many are still to be found. But be warned, do not just go off and start to dive on them without prior authorisation from the necessary authorities (this will be discussed later). Some of these sites are simply anchorages used by ancient mariners who sought shelter from adverse winds and the finds usually associated with them are anchors, ranging from the old stone anchors to the medieval iron ones, and remnants of amphoras which have been jettisoned overboard after breakage, but which can still tell us much about the mariners who had been using them.

The vast number of these roadsteads is easily understood if one considers the craft of those early days. They were minute by present-day standards, frequently measuring no more than 10–12m in length and 3–4m in beam. They were 'powered' by one square sail and later models were assisted by banks of oars and were called biremes or triremes dependent on whether there were two or three banks of oars. But even with these they were unable to progress very far against unfavourable winds. Coupled with this were the very rudimentary navigational systems employed, and as a result, the vessels tended to keep within sight of land where possible. If the wind turned foul, they beat to a safe anchorage and sheltered until the wind changed. Sometimes, of course, they did not make it and in the vicinity of most known anchorages one can often expect to find remains of wrecks. By and large these early anchorages were either islands or peninsulas which offered shelter from the winds, but of course, rivers and inlets were also likely spots.

In time, some of these roadsteads became so frequently visited that establishments or trading posts were set up, dealing with the local inhabitants of the adjoining land. These, in

The where, why, and how of underwater archaeology

turn sometimes developed into colonies. If this was not possible due to the antagonism of the locals, an island or a peninsula still offered the best defensive situation, again influencing the choice of those areas which were built upon.

Still later in time the more important of these colonies enlarged into quite big cities and really complex harbours were built. An interesting point here is the frequent omission of outer sea walls or breakwaters which sometimes confuses the diver/archaeologist who automatically assumes they are to be found. This is, of course, due to the choice of site. Where a natural land mass provides adequate shelter from the winds and where any major change of wind direction can be countered by moving the vessel to the lee side of the island or peninsula, there is no need for outer breakwaters. Nora, in Sardinia, is an example of this, while Apollonia (charted by Dr N. Flemming) is just to the contrary with extensive sea walls supplementing natural rock reefs offering protection from seawards.

The question bothering some readers by now will be—why are these sites under water? In fact they are not all under water, some are well inland and completely dried out, even cultivated by now, certainly of no interest to the diver. This is due to the fact that the sea levels are constantly changing. In England, for example, there are places where the sea has retreated; thus, Rye was once one of the larger of the Cinque Ports, but now the sea is some two or three miles away. Yet on the east coast of Suffolk, the old village of Dunwich has been inundated by the sea and now lies approximately two miles off shore. The bells of its church are still heard to ring —or so the locals would have you believe!

There are two schools of thought, much too deep for us to get into here, concerning these changes in sea levels. One holds that the earth's crust is changing due to internal movement— the Tectonic theory; and the other that the water level is rising, mainly through the melting of the ice caps—the Eustatic theory. Which of these is correct, or if as is more likely, they are both in part correct, does not concern us here. For the result is the same, and divers swimming along underwater cliffs can sometimes find evidence showing quite clearly. This will generally be in the form of deep, smooth, horizontal 'grooves' worn into the face of the cliff by wave action in centuries long gone. At times two or even three can be seen at various levels and clearly show water levels as they had been at different times.

Underwater 'beaches' are known to diving geologists, where repetitive beach levels are found in progressively deeper water. Each has been formed over a certain period of time in the past tens of thousands of years. Fortunately, the span we are interested in covers at the most, say, the last 2,500–3,000 years from the point of view of wrecks, and rather less, perhaps 1,500–2,000 years, for land installations. During this period, there appears to have been a general change in levels in the region of 1–1½m upwards. Reference to either a rise in sea levels or sinking of the land is carefully avoided, so that we do not enter into the Eustatic v Tectonic controversy!

This will mean that most of the low-lying land areas of 2,000 years ago will now be under this depth of water at the most. It means that quays, etc, which may have been built to a convenient height to load vessels, say 1m above water level —ignoring the tide which in the Mediterranean is fairly nominal—may now be under 50cm of water. Causeways linking islands to mainlands, previously perhaps a similar height above the water, will now be covered to a shallow, definitely non-diving, depth! They can usually be clearly traced by waders, assisted by prodding rods if the bottom is particularly muddy. Where the change of sea level has been gradual, the land remains are nibbled away by the sea gradually, so that they are swept away by currents and all formations are lost.

However, before we become too dejected at the thought of not finding diveable remains, it must be remembered that many parts of the Mediterranean are subject to earthquakes and where these have occurred, the land mass may have been precipitated below the sea. The earthquake itself will no doubt have done much to destroy the buildings, but at least their foundations, wall patterns and general layouts remain, as do major structures such as roads, quays, docks, canals, and harbour walls.

These land shifts also account for the fact that some of the original harbour may be high and dry on land while other parts are under water at depths much greater than the general sea level change would have caused. On the west coast of Italy, the harbour of Ancidonia is just such a case. Some of this extends for probably a kilometre inland, but a small section is off the present beach in relatively deep water. Similarly at Nora in Sardinia, the fault appears to have occurred on the east side of the Capo di Pula, the peninsula on which Nora was built. Consequently, the area to the east of the land is fairly level and under perhaps no more than 1m of water at the most. The land rises as a shallow cliff from the water's edge and, from the top of this, has a general shallow slope towards the west and continues this slope right out into the water for a distance of 1,500m reaching a depth of 3m. From this, one can trace the walls of the harbour area itself, dropping down to levels of 7–8m.

In such circumstances, not only can worthwhile remains be found which can be accurately plotted, but one can even dive! Maybe this will only be in 7–8m, but there are the advantages of good visibility and long endurance, for little air is consumed at such depths—and as has been said, it is just as wet as it would be at 40m!

3

PRELIMINARY RESEARCH

EVERY DIVER dreams of one day suddenly finding sunken treasure and of revelling in the thrill of scooping up gold coins and plate by the handful. Lurid fiction tales backed up by the, almost as fictional, tales of actual work which are cropping up fairly regularly in the press these days enhance the desire. It only needs a diver to get hold of one of the glossy magazines, especially the *National Geographical Magazine* (dare we cal that a 'glossy'?) featuring stories set in the Caribbean with pirates' loot to be dived for in ten or twenty feet of water—and he's a goner! No disrespect is meant to the magazines for they report only actual happenings, possibly embellished just a little, and certainly such finds have been made. Indeed the superb underwater photography which accompanies these articles is enough to make any diver's mouth water, especially if last week-end he was diving in the more usual conditions of cold water and bad visibility.

But, in reality, it just does not happen that way—at least never to me. A vast amount of preliminary work usually precedes the moment of glory that is recorded and this is really the secret of the successful expedition as compared to unsuccessful, pottering-around diving. Some 'homework' is essential before the physical diving part of any project is undertaken. Maybe it means turning off the television one evening a week —or if you have teenage daughters as I have, taking yourself away to another room—but it must be done. In fact, you may be surprised at the amount of interest and pleasure that can

Preliminary research

come from this preliminary work. Consider it a drudge to be skipped or omitted if you wish, but you and your group will be the losers—no one else.

You have only to read some of the reports already published to see what should be done. In particular, I would recommend the account by Sydney Wignall of his research into the *Santa Maria de la Rosa*, an Armada ship that was lost in the Blaskets in Southern Ireland. This makes excellent reading and is a thrilling story which grips the reader and engrosses him with the authenticity of the material which Mr Wignall has been able to find.

Much of this information came from Spain and involved numerous and no doubt frustrating trips to obtain it, but official reports and chronicles proved invaluable in finding the eventual location of the wreck. Mr Wignall has included copies of the actual letters from the English courts of inquiry into the sinking and their interrogation of the sole survivor, the son of the ship's pilot, who was subjected to torture, poor wretch. There is also an extract from the log of one Don Marcus de Aramburu who was in another Spanish ship which was close by at the time of the sinking and was, incidentally, almost taken down with the *Santa Maria de la Rosa*.

This is not the place to tell Mr Wignall's tale; he has done it too well, but do read it yourself and just remember that it took nearly six years to compile. I will leave it to you to judge whether the end was worthy of the means.

Starting from scratch as is likely, we must of course, decide on what our aims are. Are we wreck-searching, and if so in which area do we want to do this? Or are there some underwater relics of a long-lost building, bridge, shipyard, or even settlement on which we would like to work? Is it on the coast, in a lake, river or down a well? Or are we just looking for any interesting project so long as it is within easy reach of our home area?

Let us exclude overseas work for the moment, for this presents special problems, not the least of which are access to the site and time available. So we will take a look at what may be involved in working on a wreck-hunt on any selected part of the coastline. This puts a project within easy reach of any diving club or group as far as travelling is concerned, for wrecks abound on most coastlines, and particularly on that of Great Britain.

I would strongly recommend that in the UK as a first move you contact the Committee of Nautical Archaeology, outlining your aims or if need be, asking for suggestions as to projects that could be worked. The address is in the appendix and the Committee may be able to advise you of possible sites of the type you are seeking, that are not being worked in the area. In any case you will have put on record your interest and intention of carrying out this work. There is of course, no compulsion to do this, but the CNA has been set up to liaise with divers and to assist in underwater work. If it is not put into the picture it obviously cannot be of much assistance.

In most areas, they will be able to suggest short cuts as to possible sources of information, give you names of local archaeological societies who may be able to help and generally smooth what can be a difficult path to getting information. A comprehensive record is being compiled by the CNA of wreck sites and other interesting projects as they become known, and it could be that one of these will provide you with the project for which you are looking. Later on, as the CNA becomes more established, it is hoped that they may even be in a position to lend equipment or funds to selected projects—so make sure you get your name on the list!

As a logical first thought we can turn to the appropriate naval charts for these certainly show wrecks, but in what profusion! Bewilderingly so—one can't see the wood for trees! Many naval charts were drawn afresh at regular intervals so that old wrecks, no longer a danger to shipping, could be removed and fresh ones shown. The British Board of Trade charts, for example, are drawn year by year, listing all wrecks in the previous year, and this amounts to a bumper crop in some years. 1875 might have been termed a 'vintage year' as far as tragedy is concerned; no less than 1,100 ships were wrecked on the British coast. The recent publication of *Cornish Shipwrecks, Volume I, The South Coast*, by Richard Larn and Clive Carter lists some 270 wrecks on the south coast of Cornwall alone, and all sank within the past 150 years or so. This is an excellent book for divers searching these waters.

A high proportion of wrecks shown on naval charts will be

Preliminary research

modern and will date back mostly to the 1914–18 and 1939–45 wars. These can still make very interesting and sometimes rewarding diving. There is a wreck in the English Channel off the Newhaven–Eastbourne shore which had, amongst other things, a cargo of record players. A local diver recovered one which he has since installed in his home and which operates well. It speaks well for the wrapping—as well as the manufacture! But if you dive into or even on to these wrecks, do be very careful: they can be extremely dangerous because of the presence of sharp and jagged plates, possibly resulting from wartime action, which can be very hard to see in the dark and only too hard to feel when your head hits one. Cuts also come easy when your hands are cold and numb!

The interest in a wreck becomes greater when its name is known. Frequently this will only be learnt from the wreck itself. A ship's bell with the name engraved upon it will be little affected by the sea water and will provide the answer, so will a ship's plaque fixed to some rusting bulkhead. The nameplate off an engine with the serial number, can provide a clue which can be tracked down through the manufacturers. Many items can turn up which will tell a tale.

In 1960 the writer was diving off Panarea, in the Aeolian Islands to the north of Sicily. We were searching for a wreck which our fisherman boatman said his grandfather remembered being sunk in the vicinity. No real clue as to its location could be gained, but a reef of rocks seemed as good a starting place as any. We started cruising up and down hoping to see a shadow on the sandy bottom some 30m below which might indicate its presence. Visibility was such that we found later that at a depth as great as 40–50m we could see 30m horizontally and good colour photos were taken without flash!

We soon saw a likely shadow and there she was—on a sandy bottom 50m down. Lying quite level athwartship, but sloping down to the stern. All her wooden decks and superstructure had gone, apparently rotted away, leaving just the metal crossbeams of the deck spanning from side to side. All this metal work was rusted and encrusted with marine growth which made it almost unrecognisable. Over the bows ran two anchor chains which stretched out into the sand ahead of the ship. We surmised that she had probably struck on the reef and the skipper had let out his anchors in the hope of holding there, but eventually she had slipped off into the deep water. Above her bows there was only a pair of davits standing sentinel and behind them an old winch. Right at the stern was her wheel and binnacle overgrown with weed, and inhabited by the ugliest moray eel I have ever seen!

Gently letting out my breath, I sank slowly down through the opening below my feet, moving gently to avoid being cut by the sharp, rusty edges of the deck beams. As my eyes became accustomed to the gloom, I saw that I was standing in the engine room. It was a strange silent world of half-light, twisted pipes and fallen girders. As I looked around my eyes caught a glimpse of something whitish almost buried in the silt and slime at one side of the wreck, close against the hull side. I gently probed with a finger, stirring up a cloud of particles which seemed to hang motionless in the still water. Something seemed to move, to slip down further, and showed more white. Reaching down with my other hand, I held it steady whilst working my fingers under the edge. I was almost working in the dark due to the sediment around me which blotted out all light. My groping fingers felt a firm smooth edge, which curved sharply round. Seizing it firmly, inhaling slightly I started to rise, keeping one hand well above my head to fend off the sharp angles of iron girders.

Emerging again into lighter waters, still gently rising, the sand and silt was washing off my find in thin dark streamers, suddenly I realised I was looking at, of all things, a fish or vegetable strainer from a fine china dinner service. Virtually unmarked by its years under the sea, the glaze was still perfect and most interesting of all, a pattern showed and in it was the name of the ship—*Llanishen*—surely a Welsh name, but off the coast of Sicily? On returning to England, inquiries through Lloyds brought the answer, that she was indeed Welsh, owned by C. E. Stallybrass of Cardiff, stranded on 16 May 1885, after being in service for only ten years—'the crew were saved'; I was very pleased to read that. The manufacturers of the china strainer, Messrs Copeland, were equally interested, and publicised the find in the *Illustrated London News*. In this instance, a piece of china added so much to the wreck, changing

Preliminary research

it from just a mass of rusted iron to a known piece of history with a story of its own.

This question of identification of a wreck is of vital importance. Should the find appear to hold valuable contents for salvage, either archaeological or monetary, it is of prime necessity that it be identified. For without its identity it will be impossible to trace ownership and conclude salvage negotiations. Given the name and location of a wreck, there are many sources of information. The following is only a selection, confined to the United Kingdom.

Most lifeboat stations keep records of wrecks in their areas going back many, many years. The Receiver of Wrecks can generally be found to be most helpful and he usually has a fair fund of information although this is not always available locally. Customs records are a fruitful source of information, for as far back as the eighteenth century 'port books' were kept which recorded the movements of all shipping and wrecks in the area covered by the port in question. Today, many of the records from the smaller ports are kept at the head office of the Customs and Excise in London. It should not be forgotten that many ports that are silted up today and inoperative, such as Winchelsea and Rye, and many others in southwest England, were busy ports in years gone by and their records are still kept in London.

Not all sources go back very far. The records held by Trinity House for instance, were destroyed by fire in 1940 and they can only help with wrecks subsequent to that date. The Hydrographic Office goes back to 1913, Lloyds Register of Shipping covers as far back as 1890, but the Corporation of Lloyds has worldwide records going back 200 years. The Naval Library at Fulham covers the eighteenth century and up to date. *The Times* newspaper has records going back to 1809. Parliamentary reports in general date back to 1801 and the Parliamentary Committee Reports on Wrecks will take you back to 1770. Every wreck had its own inquiry committee to look into the whys and wherefores of the disaster and these records have been kept.

As mentioned, *Lloyds Lists* still today give details of 'overdue shipping and marine casualties' and records of these are kept right back to 1734. Sometimes one can still find some local authoritative reports on wrecks which go back even further than the national ones. For instance, in Truro Museum there is a copy of *The 16th Century Journal of the Royal Institution of Cornwall* and various wrecks are detailed in this. Another publication covering the South West is the *Exeter Papers of Economic History (SW and the Sea)* edited by H. S. Fisher and this too can be of value. Other sources include local police stations and through them one can contact the LSA (Lifesaving Apparatus Volunteer Force). This organisation has been in existence for very many years and, particularly on rocky coasts, will have attended many wrecks where their rocket lines and breeches-buoy have saved lives.

The School for Nautical Archaeology at Plymouth, England, under the sponsorship of the CNA and direction of Lt Cmdr Alan Bax is also compiling a record of wrecks and wreck authorities. Inquiries sent to either will receive a prompt reply.

In Appendix 1 there are listed the full names and addresses of these organisations and individuals who are in a position to assist in wreck identification. It must be remembered that there is no obligation on their part to render any assistance in searches, but a properly worded approach will normally produce a helpful and satisfactory reply. In many cases this will not be as full as desired, and frequently it is necessary to continue inquiries through a number of sources to obtain the full details required. In general, it will be necessary to provide the name of the vessel before anything can be done; it is not enough to be able to give the location and possible date of sinking. Also, in some cases a charge will be made for the search of records. This may be a flat-rate 'search fee' or it might be related to the time taken to carry out the search.

We have discussed the situation where a wreck has been found and identified but obviously this cannot be the case always and many occasions will arise where a wreck has been found, perhaps purely by chance, and there is nothing visible to identify it. Here the first objective is, of course, to try and identify her from documentary sources. Your search will best be commenced locally and can be guided by your rough estimate of the wreck's age. If relatively modern, then obviously the place to go is the local newspaper office. Some of these

Preliminary research

have records going back surprisingly far. The local museum and archaeological society are well worth a try, and generally an introduction from the CNA will help you in these cases.

And, of course, there is the local church. A most valuable source of information with which to identify a wreck and a grand place to look for clues to a possible wreck if you have not already found one! A walk through practically any coastal churchyard, especially in south-west England, will produce gravestones inscribed with names and details of some unfortunate casualties of some past mishap, giving dates, location, and even the name of the vessel. Prior to 1841 it was customary to bury the bodies where they were found (having removed any valuables, needless to say!) and isolated gravestones can still be found on headlands and clifftops. But in Britain the Act of 1841 made it an offence to bury the bodies in unconsecrated ground and at the same time provided for payment to the church which carried out the burial. Consequently, very detailed records were kept by the churches to ensure that they received their just dues. Timbers from wrecked ships were used in the construction of many old cottages and it may be possible to trace back from this source although it would probably be a long and difficult task. The presence of an old anchor or cannon by somebody's front gate can also be a clue to a past tragedy in the area.

Finally, on the subject of wreck identification, there may be the time when you are the first finder and identifier. Please pass your knowledge on to the appropriate authorities to add to their fund of knowledge. The Hydrographic Office (Appendix 1) are always more than happy to learn details of wrecks, including details of cargo, etc, so that their records can be enlarged.

It may be timely here to make comment on the question of law! This is really too deep a subject to be gone into in great detail here, and indeed the law is not all that clear at the time this is being written. Various efforts are being made to regularise the situation and these have been hastened by the remarkable increase in wreck activities over the past two years. The *Association, Santa Maria de la Rosa, Gerona,* and now the *Amsterdam* have all produced their problems and increased the urgency with which the matter must be tackled. Let us examine briefly how the law now stands, as it applies in the UK. (A summary of laws in force elsewhere, particularly in America, is given in Appendix 4.)

There are four types of 'wreck', namely 'Flotsam' which is matter that has floated on the sea after a ship has sunk or otherwise perished. 'Jetsam' which defines goods that have been cast upon the sea from a ship in danger of being sunk in an effort to lighten her load, but despite all the ship has perished. 'Lagan' which defines goods that have been cast into the sea prior to the sinking or perishing of a vessel, but being too heavy to float, have sunk to the bottom, and mariners, to enable them to be recovered, have marked them with a floating buoy or cork. None of these are 'wreck' so long as they remain in or on the sea, but as soon as the seas put them on to land then they become 'wreck'. These definitions were based on findings in Sir Henry Constable's Case in the year 1601! Later, the fourth category, 'Derelict' came into being and was defined as property abandoned on the seas without hope of recovery and covers property both floating and on the seabed.

In order to encourage finders to hand over such property the Admiralty salvage payments were introduced, but it is not known exactly when. Then, unless the true owner of the property appeared to claim it, in which case he was entitled to it, the goods became the property of the king in his office of Admiralty or as it was known 'Droits of the Admiralty'. Under Common Law, goods cast ashore became the property of the Crown, but if the rightful owner appeared to claim his property within a year and a day, the Crown's rights were lost. But the Common Law ruling only applied to goods found on land and not to goods found on the sea.

Apart from wreck covered by Common Law and Droits of Admiralty, there was a third category of property, namely goods found neither on land nor on sea, but between high and low water mark. Many disputes were fought over this point between the Crown and the Lord of the Manor for instance. The Merchant Shipping Act of 1846 attempted to sort out this problem and proceeded to do so in a typically legalist way, which to the layman sounds amusing, but in fact allows no legal loophole. It says . . .

that all persons whomsoever shall find, take up or be in possession of any Wreck of the sea ... which shall have been found floating or sunk at sea, or elsewhere in any tidal Waters or cast thrown or stranded upon the shore and whether the same be found above or below High water Mark and whether wholly on land or wholly in the water or partly on land or partly in the water ... shall forthwith send a report in writing to the Receiver ...

Several changes in wreck law took place until the Merchant Shipping Act of 1894 was passed, which is still in force. This did not materially alter the description of wreck and the requirements in respect of it. Looking at the law from the, possibly jaundiced, eye of the average diver one fact clearly emerges. Nearly anything found in, on, under, or adjacent to the sea is classified as wreck and under Section 518 of the Merchant Shipping Act of 1894:

any person who finds or takes possession of any wreck within the limits of the United Kingdom and who is not the owner shall as soon as possible deliver it to the Receiver of Wrecks of the district. If he fails to do so without reasonable cause he commits an offence, forfeits any claim to salvage and is liable to pay to the owner or any other person entitled to the wreck double its value ...

The Merchant Shipping Act of 1906 extended this to include property found or taken possession of outside the United Kingdom and brought within the limits of the United Kingdom. This presumably means territorial waters, but equally presumably excludes the Republic of Ireland.

Another point we might consider is that of the finder of the wreck. Providing the finder complies with all the requirements of the Act ie, hands over or reports all finds to the local Receiver of Wrecks for the district, he can lay claim to being First Salvor. It is the local officer to whom the finds must be reported. It will not do to find a wreck in Cornwall and report it in Glasgow in the hope that the news will not leak out! It will, and in the meantime the finder may have forfeited his right as First Salvor. Not only can the claim be forfeited but he can be called upon to pay the owner double the value of the material concerned. If the owner lays claim to the wreck within one year, the Receiver of Wrecks arbitrates in any agreement between finder and owner. In such cases an agreed salvage claim will be accepted and this usually does not exceed fifty per cent of the total value and will only be as high as this if salvage conditions are particularly difficult. Probably thirty per cent would be more usual. These are not laid-down figures but precedent is a good guide.

In the event of the wreck not having been claimed within a year and one day, the Receiver of Wrecks is empowered to sell the finds and will, from the proceeds, make an award to the salvor, again usually between thirty and fifty per cent. An interesting point here: such a sale must be advertised locally and at Lloyds if the value is estimated to exceed £20.

It is quite possible to buy a wreck outright, sometimes for only a nominal sum, but as always, ownership brings responsibilities as well as benefits. Should your wreck be a 100,000 ton tanker which suddenly leaks oil all over the nearby beaches, she is your ship and your responsibility. Less drastic than this, if she turns out to be a hazard to navigation, to fishermen's net and trawls or even to bathers on the beach, you may be charged with the responsibility and expense of removing the offending wreck. Even when you have bought your wreck or concluded a salvage agreement, there is nothing to stop a diver, at the time of writing this, from diving on your wreck and lifting something from it—under the same Act of 1894, providing *he* reports it to the Receiver of Wrecks. You could then end up paying him salvage! The only recourse in such a case is to go to Law and seek an injunction to stop him from touching 'your' wreck. Providing you have complied with the requirements of the Act, and have negotiated a contract with the owners (sometimes even if you have not been able to find the rightful owner at the time) you can lay claim to First Salvor, as mentioned earlier. It is then possible to seek an injunction against the rival diver. There is adequate precedent in Law to substantiate your claim, even up to the 1969 case over the *Santa Maria de la Rosa*.

However, two important points arise out of this. One, it is

not possible to seek a general injunction to stop any diver from diving on your wreck, only against a particular person or firm who has been actually doing so. So, in theory, one could be faced with a succession of such cases. As Legal Aid in such cases is unlikely, it could cost you quite a bit! Secondly, and probably more important, once a First Salvor right has been conceded to you it is essential, absolutely essential, that you make it very clear that you are 'in possession and operating' on the site. This means having equipment working on the site. In its simplest form, this need only be the basic survey lines, grids, and markers that you need for the preliminary survey and can leave permanently on the wreck. Keep accurate records, and details of your finds should be advised to all responsible authorities and even the local press. The more that is known about the site, the more likely it would be for unauthorised finds to be reported to the authorities. Do not forget to keep reporting to the Receiver of Wrecks or equivalent authority and bear in mind that the local Coastguard and Harbour Master can help keep a wary eye on the site.

Through the action of the CNA, the principle of securing the seabed leases in the vicinity of wrecks has been established. This follows on the idea of the Crown allowing leases to companies for the search for off-shore oil and gas deposits. In the case of wreck, no rights to mineral content of the seabed are awarded, but it adds weight to any preventive action that may be legally taken against interlopers trying to work on a wreck. It does not disallow other divers from swimming on, in, or around the wreck in question, but they would commit trespass if they were to remove or try to remove anything from it. The CNA will willingly advise divers seeking to secure leases, but it must be noted that only claims supported by the CNA will be entertained for leases, and such support will only be given to divers who can satisfy the CNA as to their motives. From the foregoing, it can be seen that the whole problem is very complex and is in no way akin to the laws relating to Treasure Trove on land. There any hidden find of unknown ownership is automatically the property of the Crown, but generally the finder is rewarded with the value of the find, if not with the find itself. Not so with wreck.

Another difference is that, on land, any site with archaeological value may be scheduled or can have a 'preservation order' slapped on it and no further work can be carried out until an archaeological survey or dig has been carried out; or it can be permanently preserved as an ancient monument. Not so with wreck! Well, not yet.

The attempt has been made above to cover most aspects relating to location and identification of wrecks and the legal aspects covering the finding of wreck. Let us now go back to a further aspect of the preliminary research for a wreck. Earlier it was said that Admiralty charts showed the location of a wreck. But they do a great deal more and it is vitally important that they are studied closely before commencing operations. It is also well to remember that charts are changed from time to time and this can alter features of importance. Also charts overlap in area and date span: the feature being sought will appear on one but not another. This is particularly true of charts obtained from the Hydrographic Office (wreck department). So ensure that all possible variations of charts are examined.

The chart should be studied to establish the type of bottom upon which the wreck lies, for that can influence the method of work. For instance, if it is a rocky bottom, the remains may well be hidden beneath boulders or certainly in cracks between them. There is likely to be more vegetation which will restrict visual search operations and make systematic search more difficult. One clue, concerning which more will be said later, is ballast stones. These are easily detected on a sandy bottom but they would be difficult for the inexperienced eye to see amongst general rocks on a boulder-strewn bottom. On a sandy bottom one is immediately faced with the likelihood that the majority of the wreck has sunk into or been covered over by sand. Visibility is likely to be less and will certainly be upset by the disturbance by the divers of the sand and silt. Very great care must be taken in these circumstances. Tides and currents are likely to have more effect on the diver than when he is sheltered amongst boulders. Small artifacts can work their way down very deeply into sand as indeed can hulls of vessels which are virtually intact—remember the *Amsterdam*—probably seven to eight metres in the sand.

Preliminary research

The reader need only remember standing on a beach with the surf breaking up past his feet and then recall the undertow of the water running back down the beach and dragging the sand from under his feet to realise the power of tide and current. With a large obstruction such as a hull of a vessel, there is an initial scouring effect on the upstream side which quite quickly cuts the sand from beneath the ship. As a result it gradually settles farther and farther down, until a point is reached where the water finds it easier to go over the top. When this happens the sand builds up against the upstream side and eventually covers it entirely. Frequently, as the current passes over the hull, eddies are set up on the downstream side which suck the sand away on that side and cause the hull to tilt over in that direction.

However, in many ways a sandy bottom is easier to work and instrumentation such as sub-bottom sonar, etc, can be more effectively used to locate the wreck in the first instance. Also it is possible to 'harness' the currents to help in the actual excavations. This will be discussed later.

The point is that the charts will tell you in advance what the bottom is like, the direction and strength of tides and currents, the depth, which is a major factor in determining the time available for work, and, an equally important factor, they will provide the landmarks with which you can precisely locate your find. Many a promising wreck has been 'lost' because the diver took too little trouble with his transits and could not relocate it! Check your tide tables and be sure you know the times of high and low water throughout your dive periods and also direction of currents at all states of tide. If you are buoying your wreck for survey purposes, a drop in sea level, through a change of tides, of say five metres can easily throw your positioning out by three metres. If visibility is two metres this can be a problem when you come to try and find your base point again. This particular aspect is very much more important where one is surveying a harbour installation or something of a similar nature where many points are being surveyed after buoying.

Aerial photographs can often be of value, not only for initial location of objects but for more precise definition of the shapes, etc. If it is possible to obtain these for the area in which you are working then by all means do. These can sometimes even be found in guide books! I have done this and written to the author or publisher and obtained a print of the photograph. Rephotographed and blown up into a really large enlargement details can be clearly noted; again this is of particular value on harbour works. Aerial photographs may also be obtained from aerial survey companies at quite reasonable costs; these can be very recent pictures and will be of great value.

We have spoken mostly about wrecks so far, but much that has been said applies equally to harbour works. The major difference is that most harbour surveys are likely to take place outside the UK and it then becomes essential to obtain full permission from the authorities in the country concerned. Be sure it is full permission. A letter from an archaeologist or a museum inviting you to come, may not be sufficient; check with the appropriate Embassy or Consulate as to the correct procedure to be taken. Be sure you can give them good references as to your background, both as divers and responsible people. The day when divers, other than those on holiday doing casual diving, could work on sites and wrecks is now past. Regrettably, too, many 'expeditions' have not complied with the rules and as a result expeditons in general have become discredited. Now only accredited teams with archaeologists and sponsored by their national authorities such as the CNA, Smithsonian Institution, etc will be permitted on to sites in Europe. When overseas, make contact with all interested parties and abide by the rules! If they say you must not lift artifacts, then do not—it is as simple as that.

When I first started taking my team overseas, I gave the authorities an undertaking that we would only lift sufficient artifacts for identification and dating and that these would be presented to the museum on our departure. After six years of visiting them, we still do this, but now they have learnt to appreciate our work and realise we have abided by the rules, and generally we are allowed to keep our finds, having first shown them to the authorities. In 1969 I found a beautiful jug, only 11cm high, absolutely intact, which dated back to the fifth or sixth century BC. It is a small oil jug which was used for anointing in ceremonies and had been decorated in a unique form called 'imgubbiato'. With my heart very much in

Preliminary research

my mouth, I showed this to Professor Barrecca, who is Director of Archaeology in Sardinia, suggesting that he would be interested in seeing it, but of course, would not really want it himself! I knew very well that there was not even anything really like it in his museum in Cagliari and hardly hoped to be allowed to keep it. But, I think he realised how much I wished to keep it and probably because we were playing fairly, he relented and said I could. The jug is now a very much treasured centrepiece in my own mini-museum.

4

PREPARATION

I THINK it can safely be presumed that the divers are competent, at least as divers. By that I mean that they have learnt the techniques of diving to a reasonably high standard of proficiency. Any diver not up to this standard can be a bit of a liability, particularly if conditions are not too good. A high standard of initial training is an absolute must; the organisation which carries out this training is immaterial, but the diver must have a full knowledge of diving as related to either relatively deep water or long duration underwater or a balance of both. Decompression stops procedure, buddy breathing, etc, must all be capable of being carried out without special thought. Above all, sufficient diving should have been done to enable the diver to be underwater without being consciously aware of it.

We shall be working underwater and we must be able to concentrate on the work, using breathing sets and underwater gear just as a means of getting us down to, and, of course, back from work. Diving ability is just another tool to be used. So initial training should be carried out in swimming pools if necessary (in fact initially this can be a good thing) doing jobs that will have to be done under the sea.

Take down tape measures, locate a couple of markers on the bottom of the pool and use them as triangulation points to measure odd shapes which have been cut out of plastic sheet and weighted on to the bottom. Practise tying and equally important, untying knots under water. Try fastening lifting bags

Preparation

to weights and inflate them, bringing the object to the surface. It does not really matter whether you are 3m down or 30m, the initial fastenings will be the same. Take plane tables down and practise with these. We will go into how to use all these items in Chapters 6 and 7, but the point to remember is to practise under ideal conditions until you are absolutely familiar with the equipment, then try it out in the sea and when you are fully competent, you can start work on your site —if you have located it!

One can even practise surveying from land base lines in a swimming pool. Mark out a shape on the bottom of the pool to represent the outline of a wreck or harbour. Make miniature buoys out of small plastic bottles, pieces of cork, fishing floats, anything that can be clearly seen from 10–12m. Attach these to lead weights (fishing weights will do) and use these to mark corners and major features of the subject. On the side of the bath, mark out a base line—chalk marks will suffice— and then with theodolite, sextant, or even bearing compass would be adequate for practice, take bearings of each float and log these down. Make a drawing to scale and see how well it matches the shape laid on the bottom of the pool. Chapter 7 will detail the method of doing this.

It is vitally important to learn to take measurements of fairly complicated articles and transfer these, underwater, on to paper, sketching the article at the same time. For instance, the detailed measurements of a cannon, with a good sketch can go a long way towards identifying it and then possibly the wreck. The Dutch East Indiaman *Amsterdam* recently located off Hastings was conclusively identified by the style of cannon found on her and by the design of initials stamped on them. The art of drawing under water is not as easy as it may seem. The pencil has a nasty way of floating up out of reach as soon as it is released, meanwhile the artist is either floating up himself or grovelling lower and lower into the mud and weed with lobsters trying to nibble his ear. Breathing control and weighting become quite important. Too many divers overweight themselves; try to ensure a neutral buoyancy for this work.

Even the first time of using a 30m tape underwater can be quite hilarious if the user is a bit awkward. Thirty metres is a lot of tape when it is all wound round a diver and this has happened so many times with trainee divers, that I have come to expect it from at least one man. All these techniques can be learnt well in advance of any visit to your site, so that once you are at work no time is wasted.

One thing will soon be recognised and that is that doing work under water is very much more tiring than just swimming, so training is important from that point of view. Another factor to be taken into account is that on a normal dive the diver is swimming around fairly steadily, and using all his muscles. On a work dive, he will possibly be just drawing or taking down measurements and be using far less energy; as a consequence he will get very much colder, very much more quickly. So be prepared for this: it may mean wearing extra protection, which in turn will mean adjustments to weights; also bear in mind that a correct weight at 2m in the pool, will possibly be too much at 20m.

Apart from training in diving, preparation should cover obtaining equipment. Much of this has to be purchased but a great deal can be made, and the construction of various pieces of equipment will be described. Another small point, do familiarise yourself with the metric scale. Metric measurements are used in all archaeological work, and, believe me, once one is used to it, it is a great deal easier than feet and inches.

Which leads us logically into one of the first of many items which can be made up during the winter ready for spring diving.

BASE LINE

This is very simply a measured length of rope which is divided into segments of definite lengths. The best material is a rope such as that used for clothes lines; this is rotproof polypropylene plastic core covered in a clear PVC outer case. Get the clear, non-coloured rope for this will reflect the light under water and appears white, standing out clearly for a very long way.

This should be cut to 50m and a bight put into each end. Do not try to knot this; it does not like it and will generally come undone, so whip the end into a bight having first applied a little heat from a cigarette lighter to the two adjoining faces of the rope. This will cause them to weld together and the whipping will make it very firm.

Having done this, take a roll of black PVC adhesive tape and wind this round helically, allowing at least a thirty to forty per cent overlap. Continue this for exactly 50cm, finish square and leave a gap for another 50cm before repeating the wrapping. Do this for the whole length of the rope and you will end with an alternate black and white rope, each segment measuring exactly 50cm and overall 50m long. This can be improved by taking the glass fibre type of adhesive tape, which is very strong indeed, and wrapping short lengths around the rope at every 1m point. If the 'Sellotape' strips are made long enough, 7–8cm is usually enough, you can wrap it around and stick the tails together like a flag. On this write in black paint the successive measurements. With this it becomes very simple to take off measurements for triangulation and cross-measurements which have to be done at fixed intervals. The line will appear very distinctly in photographs as well (p 54).

DRAWING BOARDS

Again there is a variety of materials that can be used, but I have found that two pieces of white Formica measuring say, 40cm × 30cm, stuck together with impact adhesive, so that the white is outwards, of course, serves very well. Drill a small hole at each corner of one short edge and thread a cord, long enough to go comfortably around your neck so that the board can be rested on your weight-belt and be suspended at the outer end by the cord. This makes a fairly firm writing platform. It can be elaborated by drilling two holes about 2cm down from the cord hole side, in which are inserted two brass or chrome-plated wing nuts and bolts. The nuts clamp a narrow crosspiece of wood, metal, or rigid plastic which secures the paper on to the board. Or you can simply be lazy like me, and just use a fairly hefty 'Bulldog' type of spring paper clip. This will rust like anything, but will easily last one season.

Take an ordinary HB pencil down with the board, and, either tie it, with say, 20cm of cord, to the top right-hand

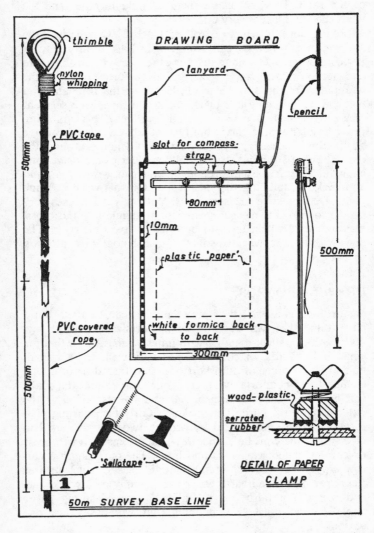

corner hole on the board or tie it to an elastic band which is in turn slipped round your wrist. I prefer the former, but it's a

Preparation

matter of personal preference. The latter method tends to let the pencil float out of reach—as you raise your hand so the pencil floats higher! We usually use one of the plastic drawing papers, preferably about 0.12mm thickness; if too thin it will wave about too much as you swim. The slightly heavier sheet sticks more firmly to the board. An HB pencil will write quite easily and clearly on this. Sharpen both ends of the pencil and not too sharp, they break easily for you cannot really tell how much pressure you are putting on to the lead, especially when you are cold and possibly wearing gloves.

An additional idea that is quite worth while, is to divide one edge of the board into black and white squares each measuring 2cm; this serves as a handy scale. A further refinement would be provision for a compass and/or depth gauge to be clipped to the top of the board. Both of these instruments are essential tools of the trade and it is certainly more convenient to have them both in front of you as you work, rather than to have to keep looking at your wrist (fig opposite).

Two words of warning here, in respect of the compass. Be sure that it is not affected by your paper clip or the buckle of your weightbelt or any other lump of ferrous metal you may be carrying, and be sure when you are using it that you hold it squarely in front of you. Too often one sees a diver glance at his wrist compass which he has vaguely lifted in front of his eyes, but which in fact is still off to one side, so that as he looks at it he is in fact looking at an angle to his real direction of travel. In such circumstances the compass bearing can easily be some twenty degrees off the actual heading.

With this sort of clip-board the diver can take down several sheets of paper and do a series of drawings. It is possible to write directly on the plastic board with an ordinary pencil, but only one set of drawings can be made and these have to be transferred to permanent records before the board can be cleaned for the next job. If loose sheets of plastic paper are used, they can be stored until required.

It is possible to make the boards in other materials, including fibreglass; so long as they do not float, it does not really matter, but the suggestions outlined have proved to be as easy a method as any.

UNDERWATER MARKERS

These fall into two categories, those that float above the marked object and are therefore easy to see from a horizontal position and those which mark at the level of the object and can clearly be seen from above.

In the first instance we are really talking about buoys or markers that float. It is well worth while putting a lot of thought and preparation into these items because they are a fundamental requirement for any form of survey. Basically anything that floats will act satisfactorily up to a point; your choice can be personal.

Polystyrene or similar expanded plastic, readily bought in sheet or slab form, will do, but is not recommended, as it is very inclined to crumble and break. Corks are fine for buoyancy but are difficult to mark; fishing floats are excellent but inclined to be expensive. Ordinary squares of fairly firm PVC sheet (p 57) will often float quite well and are easily fixed to cords, marking on them is also easy, but they have an unfortunate tendency to float edge-on at the crucial moment and the diver cannot then read the number on them. Another drawback of PVC is the truly remarkable speed in which algae, etc will grow on it! After as little as a week under water, the markers can become thickly covered with weed and slime which makes them very difficult to read.

The writer's preference is for cheap, small plastic bottles; approximately 100cc size are ideal, which can be readily purchased. They can be easily marked with any black paint of a top-coat type—undercoats do not stick very well to plastic surfaces as anyone who has tried to paint plastic drain pipes or rainwater guttering will know! I use a matt blackboard paint quite satisfactorily for all underwater markings.

Choose bottles with screw caps and of cylindrical shape. The latter point makes them easy to read from any direction, especially if the markings are put on opposite sides in large, say 6cm high, figures. The choice of screw caps means that there is an adequate projection on the neck around which to make fast the attachment line. The caps themselves can be left on or removed as desired. If they are removed, the water

Preparation

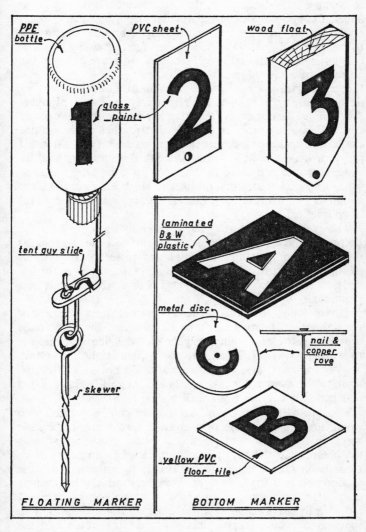

pressure will ensure that the bottles retain the full shape, whereas if the caps are left screwed tight, the bottles will tend to collapse with depth. With the caps retained, it is possible to adjust the buoyancy at any particular depth by spilling some of the air out and then screwing the cap tight; this ensures that the marker will float tightly up against the restraining line, but it is not so buoyant that it will loosen an insecure fixing to the object to be surveyed (p 57).

I am a firm believer in the use of plastics underwater; they do not rust or corrode and are reasonably easy to manipulate. Also I like plastics cords for fixing my bottles to the securing device. The only problem here, is that plastics cords are not the easiest to keep tied, especially nylon line. I use a twisted polypropylene line of immense strength that ties very well indeed; bowlines and clove hitches hold up quite safely. It is as well to try and ensure that all lines to the bottles are the same length as this helps reading from a distance and can be useful when levelling.

The method of securing the marker to the object on the bottom can again be very much a matter of taste. So long as it makes a safe fastening which is easy to secure, it does not matter—bearing in mind that in certain circumstances the fixing may have to be done on a snorkel dive when time is at a premium.

I have found that long meat skewers are quite satisfactory for lightweight floating markers. Be sure that the loop at the top has been closed completely or else the line will manage to work its way off. It helps to use a lightweight slide-clamp as used on tent guys. In other words do not make the line tight to the skewer but take the line through the skewer-head and fasten it back to itself with a slide (p 57).

For normal usage where the skewer can be pushed into wood, firm mud, or sand or wedged in a rock crack, the slide stays up the line, but at times it is better or easier to fasten the line around the object, say the neck of an amphora. Here there is no need actually to tie, simply pass the loop of line around and pull the slide tight. Time is saved, the line cannot come free, and equally important, the line can be released quite easily by slackening off the slide. For transportation, the line can be wrapped tightly around the bottle and the skewer pushed down through the windings to tie the loose end. In this way one does not have yards of loose 'knitting'

Preparation

floating around one's neck and getting tangled in everything.

SURFACE MARKER BUOYS

Obviously, the markers already described are not meant for the actual survey from land. Much larger floats are required for this, of such a shape and size that they can easily be seen through a theodolite from perhaps five or six hundred metres away.

The writer's team originally used large blocks of foam plastic with ranging poles stuck through the centre—these continually broke and were quite hopeless. Experiments were then tried using football bladders, again with no success. The action of sunlight and salt water caused a rapid degradation of the rubber, which caused the bladders to adopt most peculiar shapes before bursting!

So back to plastics! Ordinary beach balls (p 60) are quite excellent. Inflate them sufficiently to give a firm roundness, but bear in mind that the sun will heat up the air if you are in the Mediterranean for instance, and this can cause them to burst if originally inflated too hard. The fact that frequently they have brightly coloured panels makes them easy to see from afar and adds to their suitability. As with the little markers, beach balls have to be fixed securely—very securely! The displacement of a ball only 30cm diameter is in the region of 20kg, and when to this is added the jerking of the securing line due to wave action and possible submersion due to tide, it is perhaps not surprising that the float will depart from its mooring line with monotonous frequency or the line will part from its fixing on the seabed, or if secured to what has been considered to be a weighty object, it may be found many metres from its original location!

The most difficult part is securing the buoy to the line and here I have found, after many years of experimenting, the easiest and cheapest method. Go round to your local greengrocer and ask him for some of his old onion or carrot sacks! Most loose vegetables are now supplied to him in net sacks, made of plastics too! These are scrapped after emptying and you will find he is quite happy to pass them on to you. Inflate a beach ball, put it into the net bags, tie the line firmly to, and

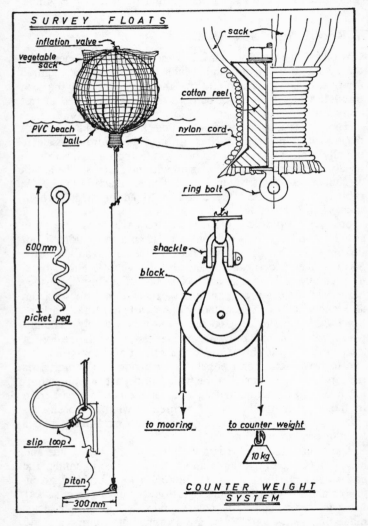

through, the neck of the sack, and the job is done. The colours of the ball can easily be seen through the netting, room for expansion of the ball is available and an easy fastening is made. This arrangement has another valuble asset: when deflated both balls and sacks take up very little room and weigh next to nothing, both major considerations when expeditions are being planned.

BUOY FASTENINGS

To fasten a marker buoy to the seabed becomes a greater problem as the buoyancy of the float is increased. Correspondingly stronger fixings have to be made, but the arrangements used on the small bottle floats, only scaled up in size, are still suitable (see fig opposite).

Make heavy iron pitons or stakes which can be hammered into rock cracks, or obtain screw-type picket posts such as used for barbed wire barricades. These can be useful in areas of sandy bottom; if screwed deeply enough they will withstand very strong pulls from above. Obviously heavy weights could be used but it is impracticable to carry these to the site. However, by using a tent guy slide fixing, a bight can be taken around any convenient boulder, lump of iron or projection, which may be on the site. But do bear in mind that these fastenings may take a tremendous beating and unless they are really secure, they may easily come adrift and all the work, not to say the buoys, will have been lost. I know, I speak from bitter experience. Almost without fail, I have found that all markers have just been nicely fixed when up blows the biggest gale for months! When it eventually subsides, markers have moved metres from their original fixing point or have vanished altogether.

Another factor plays a part in the disappearance of markers and I have been unable to devise a method of overcoming it, so suggestions would be welcome. This is the unremitting curiosity of boat-owners and aspiring long-distance swimmers who cannot resist the temptation to sail or swim out to the floats and pull them up! This can drive the most hardy and philosophical expedition leader to desperation, if not to drink. Large notices written on the buoys have no effect whatever, even when written in the local tongue. I suppose a twenty-four hour watch with a high-powered rifle is the only answer.

Where substantial differences in level due to tidal changes can be expected, the position of a buoy will vary considerably according to the state of the tide. A buoy stationed at high water with say, 15m depth of water, will move as much as 9m downtide with a sealevel drop of only 3m! So on a survey lasting three or four hours, which would not be at all unusual, even with a 6m tide there would be a swing of each buoy over some 5m with the first drop of one metre. Your final positioning could be haywire. Theoretically, these variations could be compensated mathematically, providing one had very accurate tide tables for the site area, and noted the precise time of each set of bearings, but even then wind effect would have to be taken into account as well—all far too cumbersome for our purposes.

The practical method is to ensure that there is a swimmer at the buoy when readings are about to be taken. This swimmer then takes up the slack until the buoy is vertically over the point it is marking. This is very easily done and I think it is the most satisfactory method. If a swimmer is not available, a counterweight system can be used. A block (preferably of 'Tufnol') is fixed beneath the buoy. One end of the mooring line is anchored to the seabed in the usual way, the free end is taken over the sheave of the block and fastened to another weight, which hangs below the buoy. As the tide falls, the weight is free to drop and takes up the slack in the line as it does so; as the tide rises, the buoy floats higher and uses more anchor line, pulling the weighted end up as it does so.

This can work very well, but in a choppy sea or strong tide/current there is a tendency for the buoy to pull the line until the weight is tight up against the Tufnol block. So the error still can occur. The fig on p 60 may explain the above suggestions. Problems arise in underwater archaeology as in everything else, but they are not there to stop the operation—they are there to be overcome!

This really covers floating markers and we can now consider ground markers.

BOTTOM MARKERS

These must be flat discs or squares which will not float away,

Preparation

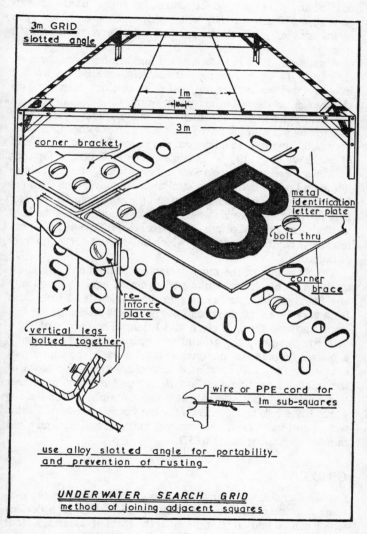

which stand out clearly from their background, and which can be marked in an easily readable fashion. They are essential when photographic coverage is being used as will be described later, for they will enable successive photographs of an area to put into the correct order in relation to each other.

Aluminium discs have been used and are quite satisfactory; they are light, do not corrode excessively, especially if washed in clean fresh water after use, and paint markings can show up clearly against them. Ordinary galvanised steel discs or squares are nearly as good, but they are heavier and more easily affected by sea water. I have found that plastics floor tiles work very well. Buy these in a light colour, yellow for preference, and of as heavy a quality as possible. They will take a top-coat paint very well, cannot corrode, and are even lighter than aluminium and much lighter than steel. Additionally, when working in an open area where there is a swell evident on the bottom, it is possible to drive a skewer or spike through them to help secure them to the bottom. Not possible with metal plates unless they have been drilled, which all makes for extra preparation.

Yellow is one of the most distinctive colours and will show up, even in black and white photography, against practically any background. Tiles will retain their colour and brightness for a very long time; the ones I use now were made up over five years ago and are still clean and bright. The choice of characters for marking the ground and floating markers is largely a matter of personal preference, but basically remember that frequently they will be related to one another. Also more of the smaller float markers than the ground markers are likely to be required. Because of this, I use the alphabet for the tile ground markers and numerals for the floating markers. In this way, I can have twenty-six ground markers and an unlimited number of floating buoys (p 57).

GRIDS

On wreck excavation and, in some cases in harbour exploration, it is necessary to use grids (p 63). These are devices which can be laid on the bottom in the form of squares, so that artifacts which are found within them can be located accurately for later recording and drawing.

Page 65 (above) The 1969 Mensura Diving Team expedition to Bithia, Sardinia. Composed of Royal Air Force divers (with the exception of the author, third from left) this was the third British service team taken abroad; *(below)* setting up a base line theodolite station at Nora, Sardinia. The tower in the background is sixteenth-century Spanish

Page 66 The author and Mr Peter Marsden working in the quicksand surrounding the *Amsterdam*. The vertical square can be seen here being held plumb from the top while the offsets of the planking are taken. This is apparently midships, and shows the hull to just above the level of the lower gun deck ports. The black and white segments on the vertical square are painted on the side away from the camera

Preparation

These can be prepared in advance and training in their use can be carried out, as suggested earlier, in swimming pools, etc. Here I reluctantly have to forgo my preference for plastics at last. It is possible to make up squares with small-bore plastic pipe, but the joints have to be solvent-welded and once made cannot be separated without breaking them. No, a better bet is angle-iron, and best of all is aluminium slotted angle, for here again one has the benefit of lightness and non-corrosion. But this is not always available, in which case ordinary steel has to be used.

Decide on the overall square size (I use 3m), and cut the lengths to fit. They are usually supplied in 10ft lengths so very little waste occurs. Paint alternate 10cm segments with black paint all round the frame and a grid is immediately established upon which any find can be located within probably one centimetre of its precise location. The main square can be subdivided into 1m or 50cm squares by stretching cords across from side to side, ensuring they are tightly stretched.

Having cut and marked the four sides of each grid, lay them one upon the other and bolt them together for ease of transport. On arriving at the site it is quite easy to bolt them into a square. It pays to use a few short lengths with the ends cut to 45 degrees, as braces across each corner. This ensures rigidity and correct right-angles, which can also be ensured by having one diagonal line of precise length which can be hooked across the grid (Pythagoras!). If the site is likely to be sloping, you will require vertical supports so that the individual legs can be adjusted to keep the grid level. The makers of angle alloy often supply small brackets which can be fastened at the corners of the grid and to which the vertical legs can be bolted on site in accordance with the requirements dictated by the slope. These brackets can be fixed as part of your preparation. Also, if the site is large enough to require a series of grids, make up strengthening brackets. Obviously the flat sides of the grids can easily be bolted together on site, but I recommend that additional bracing lengths be fastened over the joints of the non-adjoining members. If your grid is likely to be a very permanent device, use scaffolding pipes and clips. These are easily assembled and extremely strong and lasting, but too heavy for easy moving (p 68).

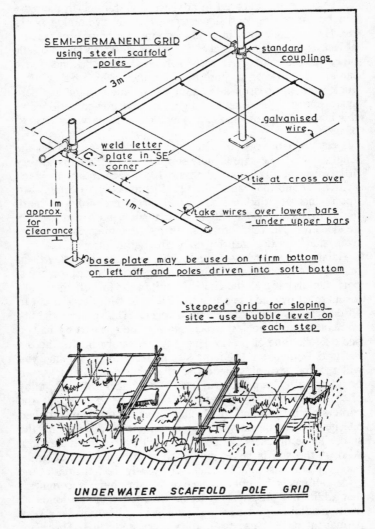

A development of the grid which puts it into a three dimensional unit, is the box frame developed by Ian Morrison. I am indebted to him for his help in compiling the following description and details of the method of operation, which are set out more fully in his own publications (see bibliography).

The principle of operation is based upon orthodox photogrammetry which depends upon precise knowledge of the internal geometry of the camera being used, with which few control measurements are required within the survey area. However, in underwater work, we are concerned with much smaller areas (due to limitations of visibility, if for no other reason), have at our disposal much less precise cameras, and can less accurately control photography. By supplying control measurements within the survey area and having these appear in all photographs taken, we can achieve the necessary degree of accuracy of measurement with simple cameras and relatively poor control of photography, the two main requirements merely being that our control measurements and objects to be surveyed are clearly visible in the photographic print.

The basic unit of equipment is a skeleton box frame of known dimensions. For preference, this should be square on plan and should consist of two square grids laid one over the other in parallel planes and separated by vertical corner supports which should not be much less than one quarter of the length of the horizontal members, although this is not too critical (p 70).

All members are then marked or calibrated into known sub-dimensions by painting alternative black and white segments. Waterproof PVC insulating tape makes a quick alternative to painting. Care must be taken to ensure that the calibration marks correspond exactly on both upper and lower members and that parallel members have the same sequence of markings. It is also of value to clearly mark the mid-points of all the horizontal members.

The materials used to make these frames can be pipework (plastics or metal) using conventional plumbing fittings, but a simpler solution is to use materials that can be easily assembled on site, especially if the site requires larger units. Square section tubular metal lengths can be obtained, complete with matching fittings, from companies supplying shop or store display units. The jointing system is a plug-fit type so it is possible to assemble the unit actually under water on site. Where this is

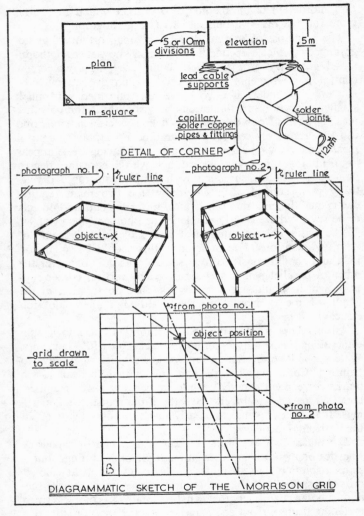

to be done, it is recommended that each fitting be firmly joined to a length of tube by an epoxy resin joint, so obviating the possibility of loss underwater and facilitating the joining together of the frame under adverse conditions.

The diver places the portable marker frame on the seabed and over the object(s) of interest. The true shape of this frame is known, and its distorted apparent shape on the photographs defines the perspective in which they show the seabed. Having positioned the frame, it is only necessary to photograph it from a minimum of two points (a third is really desirable since it allows a check to be made for errors), ensuring that both the complete frame and the object(s) being surveyed are clearly visible, and enabling an accurate position to be worked out on the photographic print at a convenient time. In practice it is worth taking ample photographs to ensure that the area is fully covered. For a plan of the area, photographs from overhead will reduce the amount of dead ground. These should be augmented by others taken looking down on the frame from an angle of approximately 45 degrees at regular intervals of a similar amount around the periphery of the frame.

Back on land, first draw a plan view of the true shape of the frame at the desired scale showing the calibration marks (this basic drawing can be used again and again if the actual plotting is done on tracing paper overlays). Then take a straight-edge and lay it across the first photograph (paper print *or* projected slide) so that it passes through the image of the point to be plotted. Pivoting it on this point, watch where the straight-edge cuts the calibration marks on the top and bottom horizontal members of the side of the box frame nearest the foreground of the photograph. Stop when it cuts through corresponding calibration marks (eg three marks right of the midpoint on the top member and the same on the bottom one). Then look at where the straight-edge cuts the image of the far side of the box. It will be found that it again cuts at marks that correspond on the top and bottom members, although the value there will be different from that on the near side of the box (eg four marks left of the corner on both top and bottom of the far side). When this has been checked, transfer the straight-edge to the plan and draw a line joining 'three marks right of centre on the near side' to 'four marks left of corner on the far side'. The true plan location of the point of interest then lies somewhere on that position line.

Repeat the procedure for the image of the same point on the second photograph and transfer the second position line in the same way. The true location of the point is where the two lines cross. A third line obtained from a third photograph provides a check.

Once the plan position of the object is known (ie its position in the plane of the base of the frame), this can readily be transferred to one of the photographs by using the straightedge across the markings of the base of the frame. The difference between this point on the photograph and the position of the image of the object represents the height of this object above the base. This height can then be read off accurately by using the calibration marks on the vertical edges of the box frame.

This method is at its most useful when a mass of detail has to be recorded in a relatively small area. Ian Morrison and others who have used the system find its operational advantages greatest in two contrasting sets of conditions. Firstly in deep diving it is much easier for a diver dulled by narcosis to drop a frame on the seabed and take a few photographs than for him to make hundreds of manual measurements. Secondly, the method allows a great deal of shallow water survey work to be done by snorkelling and this can be a considerable convenience on archaeological reconnaissance.

Ian Morrison has used this method on submerged Greek ruins and on the wreck of the *de Liefde* off Shetland, as well as in marine biological work off the Canaries. He has set out full instructions in the 1969 Underwater Association Report, together with illustrations.

It is, of course, best that the cage be level in the first instance and Ian Morrison devised the idea of using heavy, leadcoated electric cable. One end of each of four separate pieces is fastened securely to a corner on the lower frame members and the rest is formed into a wide, conical coil. These can be easily extended or compressed until a spirit level laid across the top members shows the cage to be level. The electric cable will retain whatever degree of extension you desire, and the weight of it will hold the whole cage relatively firmly in place.

In principle this idea can be operated in any size, but in practice the dimensions suggested are likely to be the largest

Preparation

which can be reasonably handled. However, it is a fine idea where there are large numbers of small artifacts to record, though it does, of course, depend upon visibility being sufficient to allow the photograph to be taken from far enough away to get the whole cage in picture.

If your work is likely to require the plotting of many items then it is well worth investing time and money in this device.

VERTICAL SQUARE

Still on the subject of using frames, a useful device can be made up, again from slotted angle. On the *Amsterdam* we were faced with taking offsets against the exposed side of the hull to establish its profile. Time was very much against us as, by the time the hole was dug, even using mechanical excavators, the tide was on the turn and the hole would soon fill again. The normal method of taking these measurements would entail dropping a plumb-line for a vertical datum and then measuring from that to the hull at fixed intervals. Two problems arise, firstly any plumb-line is inclined to swing about and either time is wasted allowing it to settle or inaccuracies occur. Secondly, one would have to measure first down the plumb-line and then from it to the hull, again time consuming. Of course, it would be possible to lay a ranging pole against the side, but a plumb-line would still be necessary to ensure it was vertical.

What I did was to take a 2.5m length of angle, calibrate it in 10cm segments and number these from zero at the top. Through the zero mark I attached another shorter length of angle at exactly right-angles, using a set-square to be sure of this and double-checking by measurement. This was held accurately by a cross-brace of angle which was bolted to both members, complete with lock-nuts. Along the short top member, I fastened an ordinary carpenter's level (p 74).

It was then only necessary to rest the horizontal member on the top of the exposed ship's timbers, or to rest the lower end on the bottom of the hole and the horizontal member against the side of the hull. In either case it was adjusted plumb by means of the spirit level and held that way, whilst another diver in the hole merely called off the distances to the hull side from

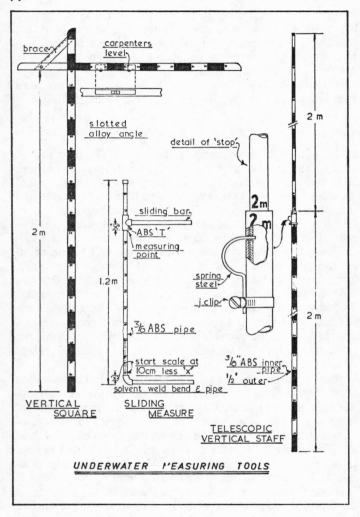

the upright, opposite selected measurements on the vertical staff. In this way, many measurements could be very quickly noted.

LAYING OUT A RIGHT-ANGLE

On the subject of right-angles, frequently it is necessary to lay a lead line at right-angles to an already established base line. This does not appear easy; obviously one cannot use a set square or T-square to achieve the correct angle, so how? Think back to your old geometry lessons and the Theorem of Pythagoras. 'The square on the hypotenuse is equal to the sum of the squares on the other two sides . . .' How does that help? Simply by remembering that a triangle with sides of 3, 4, and 5 units will always be a right-angle triangle. So mark on your base line where you wish to take off your second lead line (point A, p 76). From A take one tape measure out for 3m or whatever unit you like, in the direction you wish your line to run (point B). Measure either way along the base line for a distance of 4 of the same units (point C). From that point C, unwind the tape for 5 units and swing it round until it meets the first tape at B. That will be the spot through which your lead line must pass to be at right angles to your base line. Easy, isn't it?

MEASURING

Underwater measuring becomes a major, if not the most important, factor to be taken into consideration and the correct tools are required here as well.

For precise accuracy one should use a measuring chain, but our own experience with these has shown that they are not ideal under water. Certainly they do not stretch and will always provide accurate measurements—providing they are pulled taut and can easily be read. All too frequently we have found that they are far too easily snagged or fouled on some obstruction or tangled in seaweed, to give the shortest distance between two points needed for accuracy. We have also found that measuring chains are difficult to read accurately under water, especially to those unused to them and finally, because they can only be folded and not wound on to a drum, they become a bit of a problem to handle.

We originally used a steel tape of the type used on land, but found that the action of salt water, sand, etc quickly removed

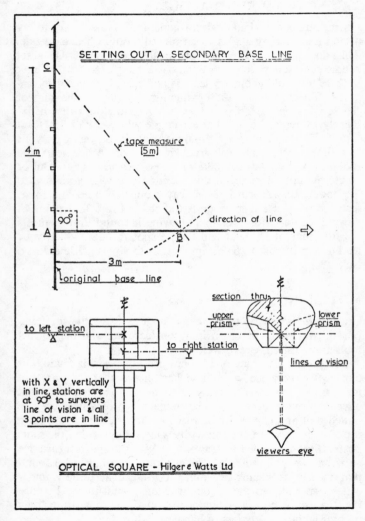

the numerals printed on it, generally within one expedition. We eventually found one which did, we think, surprisingly well to last nearly three expeditions, but eventually that too, gave up. Apart from loss of numerals, the salt water action on the rewind spring caused this to break. Additionally, as it had an enclosed case, this was prone to get clogged with sand and mud which was extremely difficult to move.

We eventually changed to fibreglass tapes and wound on open frame winders. With this type it is easy to ensure that all foreign matter is washed out from the centre of the tape and that all working parts can be oiled. The only drawback with these tapes is that the metal fastenings or clips at the end of the tape can be pulled away. Divers, struggling in the dark or against currents, frequently do not realise how much pressure or pull they are applying to a line or a tape and if the anchorage is as strong as it should be, then something has got to give.

Fibreglass tapes can stretch and if really accurate measurements are required, it pays to check your tape against a measuring chain occasionally.

So far we have endeavoured to describe some of the more simple items of equipment you must have for underwater archaeology. Their construction will provide a challenge to ingenuity and can occupy many a winter's evening. Their use in swimming pools will prepare you for some of the work on site.

5

MORE ADVANCED EQUIPMENT TO MAKE

THE TAKING of measurements alone will be of no value unless they can be related to some known point or points. Distances between two points are really quite valueless unless the position of at least one of those points is known. So obviously we must use equipment which will provide this information.

Above water we can use a sextant, a theodolite, and of course, a compass. Of the three only the compass can be used under water and it can be a highly suspect device for so much depends upon the user. Even on land it is surprising how many people cannot take a bearing accurately to within five degrees and this is equal to an error of five metres in sixty! Under water this can be multiplied by three at least, in most operators' hands, and so becomes very inadequate for accurate recordings.

However, there are devices which offer great accuracy under water and are relatively simple both to make and to use. In general, they tend to follow established land survey practice, merely being simplified for underwater use and made reasonably waterproof.

PLANE TABLE

The first of these is known as a plane table which in conditions of good visibility can give a fair degree of accuracy. It consists of two units, the first being a sighting device and the second a form of underwater drawing table. Their use will be

More advanced equipment to make

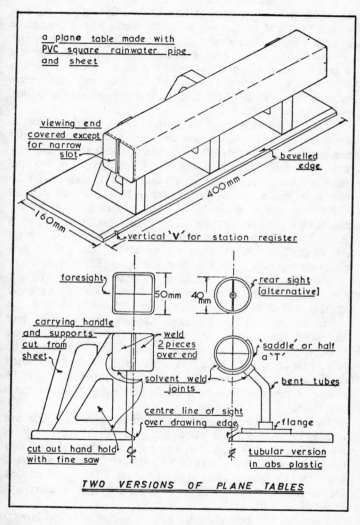

TWO VERSIONS OF PLANE TABLES

discussed later, but here are suggestions on how one can be made.

It may be made of a variety of materials, wood, plastics, or metal or possibly a combination of all. Taking the sighting device first, a sighting tube, p 79, is mounted on a base plate in such a way that one edge of the base plate is directly under the centre line of the tube and aligned with it. Consequently a pencil drawn along the edge of the base plate will mark a line exactly repeating the line of sight through the instrument, which line will become a transit line joining the station point and the object.

The sighting tube may be made from a metal or plastics pipe and may be square or circular in section. It should measure approximately 30cm long and needs to have an internal diameter of say, 5–6cm. I prefer square section and plastics. One of the square-section rainwater down pipes on the market is quite satisfactory and may be easily obtained.

At each end a graticule or hair-line arrangement for the actual sighting lines is fitted. These may consist of a straight vertical piece of piano wire cutting precisely through the centre line of the tube, crossed by a horizontal wire also passing precisely through the centre line of the tube. This arrangement may be the same at both ends or if preferred at the eye-piece end a slight variation may be used. Here one wire, vertically aligned, with a very small brass washer soldered exactly in the centre can be used as a rear sight. Or, two vertical wires may be used at the eye-piece end, parallel to each other and spaced equidistant from a vertical centre line. Looking through the tube, it is only necessary to sight the cross of the front wires on the subject, and correctly align them with the rear sight. In effect, the device is like a telescopic gunsight without the magnification.

The tube being complete, its supports and baseplate are next made. Again I use plastics; but metal or wood may be used. Plastics, however, overcomes the problem of rust and corrosion and its specific gravity ensures that it does not float, but is still reasonably light to handle. Also, all the parts needed for construction can be obtained from a plumbers' merchant or plastics distributor.

There are two materials that can be used, one is PVC (polyvinyl chloride; specific gravity 1.37) and the other is ABS (acrylonitrile butadiene styrene; specific gravity 1.06). The

More advanced equipment to make

latter is preferable as it is stronger, easier to join, and less likely to chip or split.

Whatever the material, make the base plate about 400mm long and 160mm wide. A slight bevel on the sighting edge will help in drawing. If plastic material is used, aim at not less than 10cm thickness and ensure it is perfectly flat. At what will be the rear end of the bevelled sighting edge, cut a small vertical 'V' approximately 10cm in from the end. This will serve as a locating point when you are using the plane table.

The next part is the really tricky one and that is the mounting of the tube above the base plate so that the centre line of the former is exactly over and aligned with the bevelled edge of the base plate. It is advisable to make up a simple jig for this.

The shape of the vertical supports will influence this operation. If sheet material is used the faces which will be adjacent to the bevelled edge of the base plate should not be cut absolutely vertical, but at an angle so that they project over the edge of the base plate when located in position. If pipe is used for the supports, this should be bent in gentle curves to allow the same projection.

In both cases the supports should be welded directly to the sighting tube and base plate or, if of pipe, into suitable sockets already welded to the tube and base plate. The term 'weld' has been used, but strictly speaking the process is 'solvent welding'. The jointing medium is an ABS solvent cement in which ground-up plastic has been dissolved. The solvent softens the adjoining faces of the two surfaces to be joined and they mate together into a weld or joint which makes the two parts homogeneous. Instructions for carrying out this jointing can be found on the containers and are very easily followed.

If wood is used for the base plate, be sure to allow fastenings for the weights which must be affixed to ensure it does not float away when you are not looking!

Referring now to the table itself, this is probably best made of wood for the top, with angle iron or scaffold poles for the legs, the weight of these ensuring that it stays down. The advantage of wood is that pins can be stuck into it to hold the drawing paper. On the top it pays to insert a circular bubble level so that true levelling of the table can be ensured. If plastics or metal is used for the table top, the drawing material can be affixed with metal clips of the 'Bulldog' variety, metal drawing-board clips, or even some form of Scotch tape will do quite well. The drawings on p 79 will help in the construction of this device.

COMBINED THEODOLITE FOR LAND USE AND UNDERWATER SURVEY TABLE

Assuming that a theodolite for land work is not available and that in any case an accurate sighting device for direct underwater readings is required, the following device will be of value. Needless to say the land-use part cannot be used under water, but it can be interchangeable with the underwater part.

For the land we shall require a sighting device with magnification. A telescopic gunsight will do very well and can be purchased reasonably at any gun shop or it may still be possible to get an old army gunsight. For underwater use, we require a tube similar to the one already made for the plane table, but preferably circular in section, not square.

Both are mounted, as required, on a common base, so this will be considered first. The drawing on p 86 will make the description clearer. The base consists of a non-magnetic surface or table. As it is to be used underwater, ABS will be as good a material as we can readily find. The size needs to be somewhere in the region of 40–50cm square and of a thickness adequate to provide a firm mounting for a central pivot. Probably 20–25mm will prove satisfactory.

Centrally upon this surface and with the 0–180 headings at right-angles to one of its faces, mount a 360 degree protractor. An ordinary clear plastic one will do very well, but should be as large as possible. The larger the diameter, the larger the segments of arc subtended by each degree, so making possible readings down to half and perhaps quarter degrees, even if only by estimation. The protractor should be firmly cemented with impact adhesive on to the surface of the table and additionally located by drilling and screwing.

Exactly central and precisely vertically through the base plate, a hole should be bored, a close but free-moving fit for the shaft upon which the sighting tube will be located. Plastics material being non-absorbent, is better than wood for this base;

Page 83 A high level view of the *Amsterdam*, showing magnetometer readings being taken. One of the 50m black and white base lines can be seen stretched fore and aft and even from this height can be seen clearly

Page 84 (above) A slotted angle grid lying over an area of small artifacts. The crosslines which subdivide into 1m squares can be seen, together with a floating bottle marker; *(below)* a diver holding the nozzle of the dredge while another makes fast the high pressure water pipe. The dredge is fitted with an extension tube in this picture.

More advanced equipment to make

a compass and a bubble level may be located on the top surface. The former may possibly be affected by the metal legs and the compass should be 'swung' to establish this error. In practice, the compass merely serves as an aid to orientation for normally the observer will zero the device onto a physical object.

Next make the central shaft (Class T ABS pipe) which can be threaded. A length of 15cm will allow for threading of both ends for a distance of 45mm at the top and 30mm at the lower end. I would use '1in nominal bore' pipe which gives an outer diameter of 33.4mm for the shaft. At 45mm from the shorter threaded end (lower end) solvent-weld an ABS flange plate with the flange towards the lower end. This flange will rest upon the top of the protractor when the shaft is inserted through the hole. Cut a length of plastics material in the shape of a pointer and fix it to the top surface of the flange by screwing and solvent welding. At the outer end of the pointer cut one large hole with an opening at least equal to the width of the protractor scale, and across this insert a 'hair-line'. This must line up exactly with the centre of the central pivot. Outside this and at a known distance from the centre of the pivot make a fixing for the end of a measuring tape, or better still, a swivel to which the tape may be fixed.

Into the open bottom end of the tube insert a spring loop, similar to those used for jointing frame tent legs. To this, a plumb-line can be attached for exactly locating the instrument over the base point. When the pivot has been inserted through the hole in the base plate, a backing nut and spring washer can be screwed on to hold it tightly in place, and yet allow freedom to revolve. It may prove better to use two backing nuts, one to lock the other—experiment will decide.

To the upper threaded end, screw on another backing nut and after this the branch of a threaded tee, again '1in nominal bore' size will fit. Two tees will be required; the main part of one should be cut down the longitudinal centre line so that the gunsight body may be secured to it by screw clips, and a sighting tube can be inserted through the horizontal member of the other. The sighting tube should be equipped with hair-lines as suggested for the plane-table unit.

To use the device, it should be mounted on a stand, probably

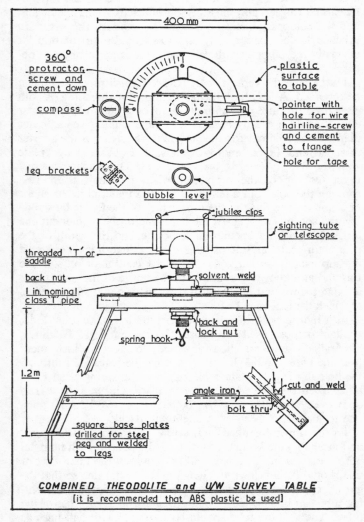

the same one made for the plane table, and secured by clamps, wing nuts and bolts, or some similar device. Depending upon

More advanced equipment to make

whether it is being used under or above water, a tee with sighting tube or gunsight will be screwed down on to the central shaft so that the longitudinal axis is directly over the pointer and then locked into place with the backing nut. By sighting through the scope or tube on to the object, a bearing can be read off the protractor under the hairline, which can be fairly estimated to a quarter of a degree.

The combined unit can be made for a few pounds and should last for a very long time.

BEARING CIRCLE AND MEASURING LINE

In conditions of bad visibility, with two divers, it will be possible to obtain bearings and distances of objects from a known point using the above instrument without sighting tube or scope. The search diver takes the tape or chain and swims off in search. When he finds an object, he pulls the tape tight—'with care!'—notes details of the find, numbers it and takes down the distance reading from the tape. His mate back at the instrument notes the bearing appearing under the hair-line on the pointer and logs it down by number. He, of course, at this point may have no idea to what it relates. However, by combining readings, we have the bearing and distance of each find in sequence. Obviously, the searcher must swim in a pattern known to the bearing reader, ie clockwise or anti-clockwise, by arrangement, so that the numbers and bearings appear in a set order. This is not an ideal method for locating finds, because the margin of error in using one bearing and a distance is too great, but it can be used when other systems cannot. With one diver only, time is wasted due to the need to return to the bearing circle to note each reading.

LEVELLING DEVICES

On very nearly any underwater project, be it wreck or harbour, one is faced with the need to obtain accurate heights of various parts of the site. These may be needed individually or related to other parts and some accurate form of measuring is required.

The general underwater gauge is far too inaccurate for this

purpose. Many do not even calibrate in less than 5ft intervals, which is obviously too great for comparative levels, where we could be concerned with differences of, perhaps, 10cm. For very shallow work, a capillary tube type of gauge will offer a better chance of estimating small differences in levels, but note the word 'estimate' for at best that is all it can be considered to be. At greater depths, even the most expensive gauge will be calibrated in 5ft intervals and many in 10ft. Bear in mind also, that all depth gauges work on the principle of water pressure at a given depth compressing the air in a capillary tube or working against a diaphragm. In effect they measure the height and weight of a column of water immediately above the instrument and record this as a measurement in feet or metres.

But this is not a constant figure, even for precisely the same spot, over a period of time. To understand this, just consider the question of tidal range. In some areas of Great Britain this can easily be 7m. So at one moment the gauge will record, say, 30m but six hours later it could read only 23m! It is easy to see how the effect of the tide could upset the calculations on a project covering several weeks or months of work. To overcome this, the exact time of every reading of the depth gauge would have to be noted. It would then be necessary to refer to a tide table covering your exact location, and also adjust the reading for the effect of the wind which was blowing at that moment, because it might be holding up, or back, the tide, etc, etc. All really not very feasible. Even using an ultra-sophisticated electronic or echo depth sounder similar calculations would be necessary to arrive at exact depths and therefore height of the site.

No, we must simplify this by relating all our measurements to one particular spot on the site and then at a convenient time accurately establish the depth of that point. This becomes the datum point for all work. Its exact depth can be obtained by a depth gauge or even a shot-line from the surface. But in fact, it will probably be found that it is not all that important to obtain too precise a reading for this spot. The question of whether a wreck is 30m or 30.05m down, is really rather academic. It is more important to ensure that the datum point is well located within the site so that all points are within a

More advanced equipment to make

reasonable distance and even within vision, where this is possible. It may be necessary to establish more than one datum point, but if so, all will relate to the original one for heights. Endeavour to establish it on a high point within the site, for this will be helpful when using some of the equipment. Of course, it may prove beneficial to locate it to one side when working on a wreck; no need to be dogmatic about this. Having established the best location of the datum point, or points, it must be marked very carefully for it may be required over the whole period of the operation.

Given good visibility the plane table or the underwater sighting device described above, can be used. This would be located over the datum point and would be a known height above it. Sight through the instrument on to a ranging staff or calibrated pole which is allowed to stand upon the spot for which a level is required. The reading which appears behind the crossed hairs of the sight will be noted and then related to the known height of the instrument as a plus or minus reading. From a series of these readings, accurate relative levels can easily be established.

CALIBRATED STAFF

Obviously one cannot take an expensive ranging staff below water, so a simple and conveniently handled unit can be made of plastics pipe (see p 74). Take two lengths, each measuring 2m and of such a diameter that one will slide within the other (1in and ¾in nominal BS 3505 pipe will fit well). On the larger pipe, starting with zero at the bottom, mark off in 10cm segments up to the full 2m. A simple white line painted around the circumference will be adequate, but painting alternative divisions in white may be clearer. It should be possible to paint the measurements clearly above each division.

Do the same with the second pipe, but start with a 2m marking some 20cm up from the bottom and continue up to 3.80m. This will allow a portion to remain within the lower pipe even when fully extended. You could, of course, make this pipe a little longer than the outer one so that a full 2m could be marked upon it, but this is a matter of choice. Obviously, the inner pipe will be free to move within the larger outer pipe so some device is needed to stop it accidentally sliding in when not intended. This can be quite an elaborate affair or as I use, merely an 'O' ring stretched and given a couple of turns around the inner pipe. As the inner pipe is raised, just roll the ring down so that it is always resting on the top of the outer pipe and they cannot telescope.

Use white polyurethane topcoat paint only, not undercoat which does not adhere well to plastics, allow it to dry thoroughly, make sure there are no burrs on the open end of the pipe to scrape the paint off and you will have a device which will stand up to a lot of wear. Also it will have almost neutral buoyancy under water and will weigh very little on land.

AQUA-LEVEL (p 93)

When visibility is too bad to allow the use of a plane table but levels still have to be taken, we have to look elsewhere for our equipment. The answer is found in the principle of the 'aqua-level' used on building sites, which depends on the time-honoured fact that water always seeks its own level. The land equipment consists merely of a length of tube, which is filled with water. One end is held level with a datum point and the other, be it five, fifty, or five hundred metres away, will be at exactly the same level when water starts to spill out of the open end. In effect we have a giant 'U' tube filled with water.

Down below, we simply reverse the procedure—we fill the tube with air! The tube in this case is of transparent plastics material, probably polythene, so that the air-to-water interface is visible. In theory, it is only necessary to use the tube as an inverted 'U' tube full of air and, as the pressure of water will be acting on both open ends of the tube, the air will get compressed equally and both air-to-water interfaces will always be level. We keep one end fixed at the datum level and swim around with the other end of the tube (even if we cannot still see the original point) and by holding a measuring rod up against the air line, read off the difference in height at the air/water interface. As with most theories, unfortunate things happen when an attempt is made to put this into practice. To start with, you can take down a tube measuring 30m in length thinking this will give you a radius of operation of a similar distance,

More advanced equipment to make

but no such luck. By the time you have got down, say, 10m, water will have pushed up inside both open ends of the tube, compressing the air. You will have to pull both ends down probably at least 3m before you get the air/water level at the correct height of your datum point. This assumes you have pulled both ends down evenly; if you have not, you will end up with a tube full of water, and will have to start all over again.

One writer, admittedly not writing from practical experience, has suggested that one can blow air into the tube. This is, frankly, impossible. It has been tried, and even if a diver were to take in a breath at one level and then rise up a few feet so that there should be a differential in ambient pressures, he would still not be able to blow air into a tube with a bore of say, 8–10mm, which is the largest likely to be met with, and in fact as large as is needed. Underwater magnification gives sufficient enlargement to make it easy to read.

So we have to devise a means of getting air into the tube to add to that already there and to force the water back so that the air/water interface is reasonably near the ends of the tube and there are no excessive lengths of tube dangling all around the place. (The tube between the ends fortunately will float up out of the way and presents no hazard.) The best way to accomplish this is to put a fairly large reservoir at the datum-point end of the tube. This can be filled with air from a small bottle or, at a pinch, can be filled from a breathing tube mouthpiece. The former is preferable as more precise amounts can be fed into the reservoir.

Ideally, the reservoir should be transparent so that the air/water interface is visible within it; after all this is needed so that it can be aligned with the datum level. The perfect solution is a clear plastics tube with a bore of at least 10cm and perhaps as big as 15cm, with an overall length of say, 30cm. This should have an open base but the top needs to be sealed in some way with just a small outlet, to which is attached the plastics tube. If Perspex is used, a small hole can be drilled in the top and a piece of plastics or copper tube inserted. The former can be cemented into place, the latter may need to be threaded and secured with nuts top and bottom, plus washers. A successful metal-to-plastics joint can be made by drilling the hole in the Perspex slightly smaller than the depth of the thread on the pipe and then sealing the bottom with a strip of adhesive plaster. Pour in a small amount of cement and leave for just long enough to soften the walls of the drilled hole and then screw the metal fitting gently into the softened hole. When this has dried it will be quite firm and strong. But the Perspex should be at least 6–7mm thick for safety (p 93). In any event, this thickness is worthwhile as it allows of rebating the edge to take the tube, thus making a stronger joint altogether.

If money or facilities do not run to Perspex tube, a quite reasonable job can be achieved by using plastics bottles such as those in which wine and vinegar are sold. With one of these, cut out the bottom; you will find there is generally a thickening near the sides of the bottle and if you cut just inside this, a strong bottom 'frame' will result. The tops are usually sealed with screw caps, so it is easy to drill a hole in one of these and insert a metal connector, screwed firmly in place with adequate washers to prevent the air from escaping.

In either case you have a transparent or at least translucent container: on the outside paint a thin black line all round. This will provide a level at which to try and keep the air/water interface at datum.

Obviously, this device full of air will take some holding down. Do this by making a harness of straps or even string, which will securely hold the container and at the bottom fasten sufficient weights to hold it firmly down. In the case of the tubular unit, a better and neater job can be made with a large 'Jubilee' clip around the tube and an iron stake, which can be driven into the seabed. Fasten a small air bottle with another 'Jubilee' clip just below the reservoir, with a piece of pipe bent up into the base of it. A quick turn of the valve and air can be let in.

You will find that a very small change in the water line in the reservoir will make a substantial change in the level in the other end of the comparatively small bore plastic tube. It is, of course, quite possible to purge all water out of the tube by feeding air in until bubbles come out of the far end. You may find that as you swim with the open end of the tube, water will enter and will force the air back. If this is not watched for, the

More advanced equipment to make

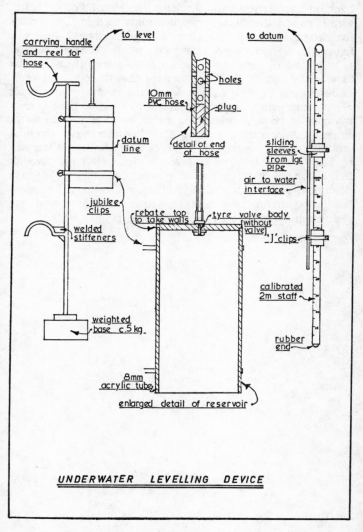

UNDERWATER LEVELLING DEVICE

water level will recede quite a long way up the tube and you will be in trouble. This can be minimised by bending the tube back on itself as you swim, so that it blocks itself off. A better way is to use the chromium-plated clips used in laboratories for cutting off flows through rubber tubes, though they will have to be pretty strong to squeeze the plastics tube sufficiently. It may even be necessary to add a small piece of rubber tube as an extension to the PVC, just so that the clip will work more efficiently. With such a device the swimmer can forget about water getting into the tube, and by pressing the clip only when he is taking readings, no air is lost, nor yet can water enter. Another alternative is to plug the open end of the tube and then drill a few holes in it just behind the plug. Water will be free to enter here under pressure change but not through the action of moving the tube when swimming. We will describe the actual method of levelling in Chapter 7. But again this is something you can make well in advance of your expedition and practise using in a swimming pool.

6
SEARCH TECHNIQUES

THE HOMEWORK has all been done, the equipment made and its use practised, and eventually the day arrives when we are standing on a beach or a cliff top looking at a frighteningly large area of sea. If we are lucky it will be calm and placid, but more likely, it will be whipped into a froth of wave and spume. As we look, not saying much aloud, we are probably all saying to ourselves, 'Oh gawd, where in hell do we start?' Well, let me assure you that wherever we start will probably be the farthermost point from your eventual find? It just happens that way.

This is where the homework begins to pay off. Study of charts will have told much about the bottom, about the tides and currents in the area, where there are races or dangerous rocks and, of course, the depth, which will play an important part in determining the method of search.

Firstly, let us assume we have ideal conditions, that is, relatively shallow water and good visibility. It does happen sometimes, which is really the reason why I have done most of my work in the Mediterranean! Whether we are searching for a harbour or a wreck, it is unlikely that we shall suddenly find it lying with a great arrow pointing to it labelled 'wreck' or 'harbour'! No, in either case the clues will initially be small and probably very hard to see or at least to recognise. So it is essential that every square metre of the seabed is visually searched carefully.

SWIM-LINE SEARCH (p 97)

For many years the author's team has used what is now known as a 'swim-line' search and in its simplest form this consists of a line of snorkel swimmers swimming side by side, looking at the seabed they are passing over. The thing is to ensure that they swim in a straight line, that they maintain a constant distance apart, and that this distance is no more than about two-thirds of the width of their extent of vision on the seabed. This will vary according to the depth of water in which they are swimming. Assuming a depth of say, 5m and a horizontal visibility of 7m at the bottom, it would be safe to allow about 5m between swimmers, but as the water became shallower or visibility reduced it might be necessary for them to close up. Ideally the arc of vision should be kept within 45 degrees of the vertical.

If the search is in a bay between two headlands, I find it best to start the line at right angles to the beach and at the foot of one headland. In this way, each swimmer snorkelling on the surface, can pick himself a landmark on the far headland and by taking an occasional glance at it as he swims, keep himself on course. Each swimmer should take his pace from the centre man, who should be, if possible, a little more experienced and capable not only of keeping a straight course himself, but also of keeping a wary eye on the others to stop them bunching. Swimmers should occasionally check with their lead man, in case of trouble. It is usually possible to communicate with each member by shouting, as all are on the surface.

If there are no headlands and one is on the open coast, I believe that in the interests of accuracy and safety, it is better for the swimmers to enter the water from a boat cruising parallel to the shore, and take their marks off features on the shore. In this case, especially if a large area has to be swept, it pays to put ranging poles on the beach in pairs, one pole at least 5m behind the other, at the extremities of the search line. Both end men can use them as transits and swim straight towards them. If the first sweep is unsuccessful, then the first set of ranging poles is moved along the beach past the second set

Search techniques

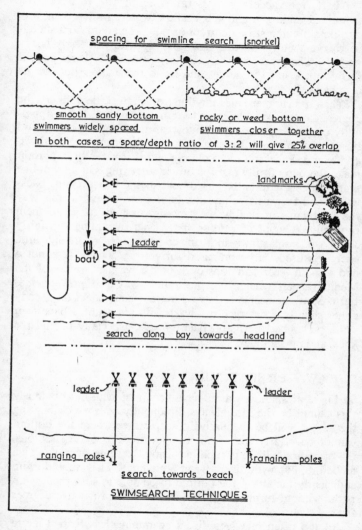

and forms another transit mark for the next stage of the sweep.

Apart from accuracy, it is rather more safe to have tired swimmers ending up in the shallow water, rather than way out, possibly in a strong current or tide. The guard boat should be carrying buoys, in this case possibly with just fairly heavy weights. The greengrocer's sacks come in handy here; just fill them with a handy boulder and tie to the marker buoy in its own bag.

When a swimmer spies anything of interest, he calls up the guard boat, snorkels down and examines his find. If it is apparently of interest, a weighted buoy can be dropped on to it and the swim continued. Later a pair of lung divers can go down and examine the find more closely and if it is really important make the buoy firmly fast for future surveying. On a swim-line search of this nature, the guard boat should be traversing backwards and forwards behind the line of swimmers; in this way the boatman can readily see any signals and attend to them promptly. A word of caution here. Line swimming is a fairly slow, gentle pastime and, in consequence, relatively little energy is used, so be prepared for getting chilled. It is not a bad idea to wear suit tops even in the warmer waters of the Mediterranean where this precaution will also prevent sunburn. Sunshine seems to have a more drastic effect on a body either slightly submerged or awash with sea water. I have seen some pretty bad burns on back of legs, especially behind the knee, and on shoulders.

DEEP WATER SEARCHING

So far clear shallow waters have been assumed, but obviously this cannot be the case always. Frequently with wreck hunting the search will be in relatively deep water where the bottom cannot be clearly seen from the surface. In such conditions, the operation will be carried out using the same swim-line idea as before, but with divers using aqualungs. They should swim sufficiently far above the bottom to be able to scan their swim paths without turning their heads more than forty-five degrees from the vertical.

All too often one sees divers swimming less than a metre above the seabed and zigzagging round boulders etc. This is no good for a sweep search. No one can see through a boulder,

Search techniques

and it could be on the other side that the clue may be lying. The golden rule is to swim as high above the bottom as possible, consistent with being able clearly to see the bottom on the extremities of the range of vision. In other words, it is of no value to swim 1m up with a horizontal visibility of 10m; being able to see horizontally does not help you to see the sea bed! Equally it is of little value to swim 5m up with the visibility only a maximum of 4m. You will not even see the bottom immediately under you, let alone at either side.

Once again visibility will automatically dictate the spacing between swimmers (or divers in this case), but the type of bottom will also make a difference. Over a clear sandy bottom divers can space out to the maximum lateral visibility, but over a very rocky bottom covered with boulders and gullies, they must swim more closely together so that they are looking down more directly into each gully.

In deep water there is no question of individual divers being able to check their own headings against landmarks and the very best and most experienced will soon be wandering off course. Even with compasses, it will be difficult to maintain a given path, a given heading, yes, but it is possible to be on the correct heading yet 2m to right or left of the intended path, so leaving a gap between two divers. In that gap could be the significant clue. So mechanical guidance is needed. This can be simply achieved by knotting or flag-marking a long line with the spacing required between divers (the black and white measuring line can be used here). Each diver swims holding on to the rope behind his mark. A problem arises, however, when a diver sees something of interest. If he stops, he gets behind the rest on the rope and in endeavouring to catch up, will in all probability swim too fast to thoroughly examine the bottom over which he is passing. The solution is this: a diver noting something of interest gives a jerk signal on his rope to both adjacent men, who relay the signal on. For example, two jerks mean 'hold up a minute' and a single jerk means 'OK carry on'.

Communication underwater is one of the biggest handicaps divers have and in a swim-line search, it becomes a major problem. The fact that sound carries well under water can be utilised. If each diver has a metal probe with him, it would only be necessary for him to strike it with a diving knife to make a ringing noise that will carry a remarkably long way. He can also tap his aqualung with either a knife or probe and get an even louder sound. On hearing the sound all divers would stop and wait for another signal to tell them it is OK to recommence the swim. On the surface as already noted, each diver can see his own landmarks, and the leader can control the pace and direction of the sweep by visual reference to his own landmarks. Under water this is not possible and the dive leaders themselves need guidance.

This is probably best contrived by the end divers following compass headings and towing light-weight floats which can be easily seen by boatmen in two boats, each being concerned with one float and one diver. The boatmen can relate the passage of the floating marker to landmarks or transits, and if they are seen to be going off course, the diver can be told to turn left or right by a system of tugs on the buoy line. He then signals the turn to his associated divers by pulling the rope in the required direction. Obviously this system has a lot of problems, and very considerable experience and practice is required before the divers can accurately cover the bottom in a thorough search. Also care must be exercised to avoid catching the line on rocks and weeds.

A tremendous aid here is the boat-to-diver type of underwater communication system. At least the boatman can instruct the diver even if he cannot hear the reply—just as well sometimes! The effectiveness of this depends not only on the equipment used but also the depth of the divers. With this type of search Syd Wignall's men, in their search for the *Santa Maria de la Rosa*, under the leadership of Lt-Comdr John Gratton OBE, RN, were able to cover several million square yards of seabed. This was probably the most extensive underwater search carried out up to that time, and its success depended on excellent teamwork and team discipline—not to mention self-discipline, which is a major necessity for divers at any time.

A variation of this system aimed at more accurate coverage of the search area, in that, in theory at least, successive sweeps over the seabed can be accomplished with the minimum of overlaps while ensuring that the whole area is covered, is as

Search techniques

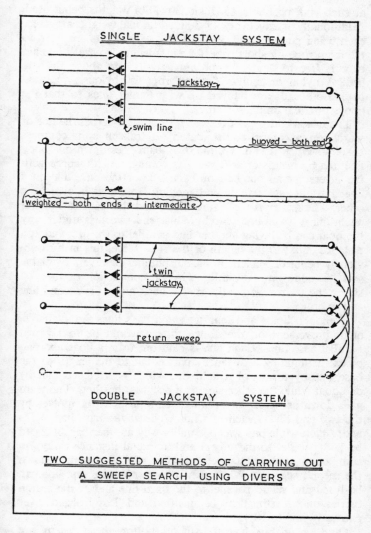

follows (see fig above). A rope is laid across the narrow dimension of an area to be searched. The rope is heavily weighted at each end, probably 15–20kg (30–40lb) weights being used. Additionally at, say, every 25–30m additional weights are fixed, each being perhaps only 2–3kg (5lb). All weights are fixed to the main rope by short ropes of equal length, so that the main rope is free to float at approximately 1–1½m above the seabed. A nylon or similar 'floating' rope is obviously required here. Both ends are buoyed up with lighter lines so that the buoys can float on the surface.

The line is laid from a boat as it progresses across the line of search on a careful course based on transits or a compass heading. When it is down, the dive team goes down one buoy line and by means of a lighter line commences its search with the divers spread out half on each side of the base line, in a line at right-angles to it and moving from one buoy to the other. The leader is the centre diver, and the team carries out an ordinary swim-line search. While one sweep is under way the boat can be laying another line parallel to the first and at a distance equal to the length of the sweep line, this in turn obviously being dependent upon the number of divers available and the underwater visibility. On the completion of the first sweep the team reverses the search back along the second line and so on.

If enough boats and long weighted base lines are available, an improvement on the above system would be to lay two parallel base lines. Each line is paid out from a boat, the two boats maintaining a distance apart equal to the length of the base line by means of a light line which is stretched taut by one boat while the other steers an accurate heading. The team goes down as before but the members position themselves by the two end divers each of whom follows the base line at his end. Again while one sweep is under way another base line can be laid parallel to the others and an equal distance away; on completion of the first sweep the line of divers returns down the second segment. While this is happening the first line can be lifted and relaid parallel to the third line and so the search carries on with the lines being lifted and 'leap-frogged' over two each time (p 101).

On a reasonably clear and smooth bottom these systems are ideal, but on a rocky or heavily weeded bottom, laying the lines tautly and lifting cleanly are both difficult tasks. Also the extra gear necessary is an additional expenditure which may only be justified on a large expedition.

Harking back to the original swim-line technique, instead of the end divers towing buoys (in itself a tiring job) which are monitored by the attendant boats for directional accuracy, it is possible for the boats themselves to tow weighted lines while keeping accurate courses on transits or bearings. They must make due allowance for effect of wind and tide and must travel very slowly so that the end divers can easily keep abreast of the weights, which they simply follow.

In practice this method can only be used in calm water conditions and with good underwater visibility. For not only must the boats be able to move slowly and still maintain accurate headings, but also they must be able to watch for signals from the lead divers, in case a sudden stop is necessitated through, perhaps, the rope becoming fouled around some obstruction. Absence of swell is also a prerequisite of this type of operation, otherwise the lead divers will either be going up and down like a yo-yo, or will have the slightly traumatic experience of a large weight whistling up and down in their immediate vicinity. But given ideal conditions, this system works.

If it is likely that your project will call for this type of search, I cannot stress too highly the need for extensive practice. Once the area has been covered it is obviously necessary that it may be considered accurately and fully searched. If this is not the case a vital clue may have been passed by and will be lost for ever.

As it is necessary to be able to mark all 'likely' finds for relocation and closer examination, some form of markers must be available to the divers. In the case of the deep water we are now discussing, the diver cannot call up his attendant boat and ask for a marker float. He must carry them with him and of necessity they must be small and not too buoyant.

DIVERS' FLOATS (p 104)

A wide variety of types may be made. Ideally they should be self-unwinding, so that as they are released they will float to the surface unwinding their mooring line as they go. If the reader imagines the old 'yo-yo', he can get a good idea of what

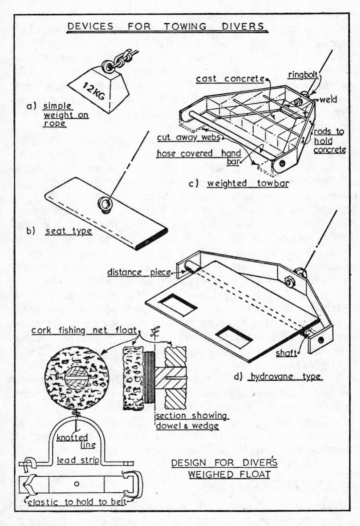

is needed. The line must be very light and should be wound round the body of the float. The free end should be fastened to a lead weight and secured to the main body of the float with an elastic band. If this is of fairly heavy elastic, it may be passed behind the weight belt so that the float and lead weight are secured to the outside of the belt. It is important that the float is evenly balanced; for instance, it is not a good idea to wind the line round a wooden bar, for in all probability, the line would unwind from one end and send the bar floating upwards at an angle. This will develop a spiralling effect and more than likely the line will fall free in a snagging coil or will catch on itself and not allow the float to reach the surface.

A simple device may be made from the fisherman's ring-type of net float, usually measuring a couple of inches across with a small hole in the centre. Securely glue a wooden bar through the holes of two of these so that they are only say, 12mm apart. Make sure the end of the line is tightly tied to the central bar and wind it neatly and fairly tightly. Shape the lead weight from a narrow strip, as shown opposite, so that it will fit between both corks and around the wound line, pass the elastic through a hole drilled in one end, this in turn behind the weight belt and loop round the free end of the lead weight. This will tightly and neatly secure both line and float. Be sure the weight is heavy enough to anchor the float firmly!

On sighting an object worth further investigation, it is only necessary to release the elastic band. The weight will drop to the bottom and the line will unwind as the float races to the surface. It is well to paint the floats with an orange fluorescent paint, for ease of relocation. It is possible to buy these buoys, professionally made, which are rather more refined but no more effective.

UNDERWATER SLEDGES (p 106)

Another system of search which has been reported is that of towing a diver behind the boat on an underwater sledge. This can certainly be used very effectively where time and divers are limited, where area is vast, where visibility is pretty good, and the subject is reasonably large.

The sledge should be made as light as possible but of non-buoyant materials for preference, although this is not essential. Basically it consists of a frame or surface upon which the

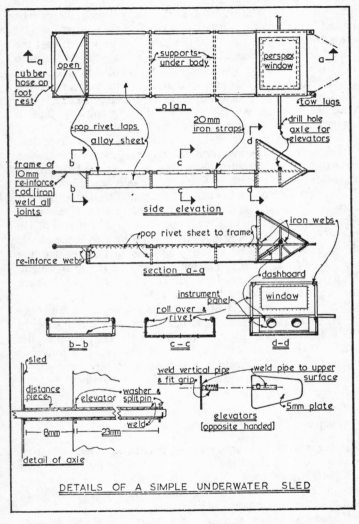

DETAILS OF A SIMPLE UNDERWATER SLED

diver lies, and should have means of altering height and direction. This is usually achieved by locating ailerons or vanes

Search techniques

near the front of the unit, certainly ahead of the centre of gravity, each of which can move independently of the other. The whole contraption is secured to a line or rope by means of a swivel hook or eye and is towed at the end of a line aimed at reaching the bottom of the sea some 50m behind the boat. Too short a line will limit the diver's ability to control the vehicle.

Control is effected by the diver holding on to the aileron control levers and by moving these together he can alter the attitude in a longitudinal plane and by moving them individually (or contrary to each other) can bank the vehicle in the direction he wishes to turn. It is a good idea to apply elastic to the control levers so that, when released, they will return to a neutral position. A clear Perspex 'windscreen' in front of the diver is well worth installing as, even at low speed, the drag of water on his body will prove very tiring. Further refinements could be depth gauge, compass, and watch built into the windscreen, although the latter is not really important as time control can be exercised by the boat.

These sledges are great fun to use, but in practice it sometimes requires so great a speed to keep them controllable that the ground is covered too quickly for accurate searching. By this I mean that if the boat goes too slowly, there is nothing the diver can do to prevent the sledge from settling on the bottom. He must have sufficient forward speed to enable him to manoeuvre easily, and if the sledge is too heavy this speed will inevitably be too fast for the job of searching. Remember, 1 knot equals approximately 31m per minute or 0.5m per second. In 3m visibility this allows the diver 6 seconds to assimilate what he sees. Considerable skill is certainly required for this work as the diver has to keep a very wary eye open for trouble ahead as well as carrying out his search. On one occasion one of our lads went slap-bang into a mud bank, which was hilarious to the divers who were watching the display, but nearly disastrous to the securing cleat on the towing boat, which was almost pulled from its socket. It was not exactly good for the morale of the diver who sank up to his knees in the mud—head first!

On a more serious note, a sledge is one of the finest ways of getting an embolism! If the craft is going fairly fast and the vanes have been made of such a size that reaction to their movement is very quick, the sledge can oscillate up and down over 3–5m either way in a matter of seconds, certainly more quickly than the normal breathing rate. As a consequence, should the diver happen to have taken a breath at the lower level, he might find himself 5–10m higher before he has had a chance to exhale, with possibly serious results.

In Sardinia the writer used a sledge which had been made by Canadian diving friends stationed out there, to facilitate a search for a Starfighter jet that had crashed into the sea nearby. This was made from steel reinforcing rod of about 1cm diameter, welded into a rigid shape like a stretcher. The area on which the diver's trunk rested was made from thin light alloy sheet, wrapped round the two longitudinal members and pop-riveted to itself. The space from the thighs down was left open with a cross-bracing at the tail end and another about 25cm further forward. These were used by the diver to rest his feet against to take the strain of the water pressure; without this he would need to hold on by hand and would be unable to work the aileron controls.

The ailerons were made from flat sheets of steel 5mm thick and measuring 30cm fore-and-aft and 60cm wide. A length of steel pipe was welded just forward of the centre line of each sheet and was of a size to accept a substantial steel rod which was in turn welded strongly to the main frame. This then became the axle around which the vanes could rotate. A short length of steel pipe was welded to the inside edge of the vane and was fitted with cycle-type handlebar grips of plastics material (rubber does not last long in sea water) to make more positive grips for the diver.

The windscreen, visor, or 'dodger' (nomenclature varies) consisted of a flat sheet of 5mm 'Perspex' which had been riveted to a frame made from angle iron (aluminium would be as good), so that the front face sloped back at an angle of approximately 60 degrees and was supported by infill pieces at the sides. The under part of the sledge at the front end was also filled in with alloy sheet as a continuation of the sheet supporting the diver's body. A small instrument panel was easily installed. The accompanying drawings will help to explain the layout, should the reader wish to make one (p 106).

TOWLINES (p 104)

In practice, a much more satisfactory method of being towed, for searching purposes, is to sling a heavy weight on a rope below and behind the boat. This needs to weigh some 5–10kg (10–20lb) and should be fastened beneath a short length of plank, rather like a swing seat, which is hooked with a swivel socket to the tow line. Connect the seat with a single central rope, not a bight. With the former the diver can sit astride the tow rope and can easily get on or off, whereas with a bight he would have to get his legs through the bight and would have a correspondingly difficult time getting off in a hurry.

With a towed weight, the length of tow line should be sufficient to clear the higher points on the seabed over which you are likely to be towing and that is all. In other words, the tow line will to all intents and purposes be vertical, for the weight and the diver will be a dead weight and will have no hydro-dynamic properties. A slow speed of no more than 2–3 knots is all that is required. A secondary light line connecting diver to boatman can be used as a signal line and a code can easily be devised for stopping, starting, and changing direction. For with this equipment, the diver will have very little directional control. With the sledge he could easily veer to one side or the other over a matter of several metres, and similarly he could alter his depth to follow the bottom contours, but with a weight this would not be possible. The boatman must do this for the diver on his (the diver's) instructions.

However, this method of being towed has the advantages of simplicity, no equipment has to be made, no skill is required to manipulate it underwater, slower speeds across the bottom are possible, and the full attention of the diver can be devoted to his search. On balance, I favour this method over the sledge. An advantage with both is that navigation is carried out by the boatman who must make allowances for the tide, wind, and current in the relative position of his boat and the diver he is towing.

The methods discussed so far have been concerned with large areas of water but we must also consider more localised

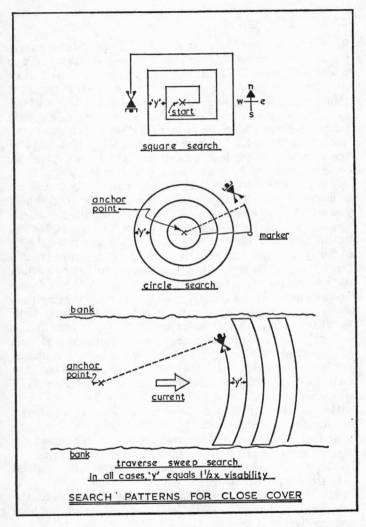

SEARCH PATTERNS FOR CLOSE COVER

search techniques which can be used either after having pin-pointed the main area of search within narrow limits or

Search techniques

where, having found our primary target, we are concentrating on a detailed examination of the seabed.

SQUARE SEARCH (opposite)

A 'square search' system uses the principle of starting from a central point and gradually working outward. In practice, the procedure on land would be something like this: start off by walking one unit of distance on a given heading, turn 90 degrees left, walk another unit, turn 90 degrees left, walk two units this time, turn 90 degrees left, walk another two units, turn 90 degrees left and carry on by increasing the length of every third leg by an amount equal to that of the first dimension. Obviously this is far simpler on land, but it can be done under water.

In the first place the unit of distance is based on the extent of the bottom that can be clearly seen and allows for sufficient overlap to ensure that all surfaces can be closely examined. An estimate of distance can be arrived at by counting leg strokes, but this is liable to be very inaccurate, for it cannot take into account the strength of tide or current. A precise measurement can be obtained by using a long underwater tape and pegging each change of direction. This limits the initial search to approximately 40 square metres assuming a 50m line is used. Naturally the diver can restart from the point at which he left off or move the start point along the axis of the search area for twice the maximum distance covered from the original starting point, and recommence from there. This is an elaborate method of searching but does have the great benefit of enabling a really detailed search to be carried out without the possibility of missing vital areas. It would be of major use when searching over uneven bottoms.

SWEEP OR CIRCULAR SEARCH (opposite)

A 'sweep' search also utilises a rope to limit movement at any particular time and is of special use where the seabed is relatively smooth but where it is still necessary to ensure a detailed examination, perhaps in the hope of finding very small artifacts.

Start by very firmly fastening an anchor point into the bottom. This can be a steel rod well driven into the sand or it could be a cuphook (2in) screwed into a hole drilled into a rock. A conventional, but large, 'Rawlplug' drill and appropriate plug will do quite well here. Alternatively, a larger hole can be drilled with a star drill and fitted with well-beaten-in lead, into which the hook is screwed. Whatever the fixing, it must be strong.

On to this hook a line, again preferably marked in distances. Commence the search by holding the bulk of the line, tape or what-have-you in your left hand and let out sufficient to allow you to swim away from the fastening but still retain it in clear view. This could be one metre or three, depending entirely on visibility and your height above the bottom. Having done this, swim in an anti-clockwise direction keeping the anchor line taut. Should you be left-handed, you may find it easier to hold the rope in your right hand, and swim clockwise. The idea, of course, is to leave your best hand free for picking up articles or using a knife while not allowing the rope to foul your body.

Note by some bottom mark where you have started and on reaching this after the first lap, let out the line until you are covering a fresh swathe of seabed, but can still clearly see your original swathe. Continue your swim in the same direction as the first lap, noting your start point once again. On reaching this repeat the operation, and so on until the end of the line has been reached. With this form of search, on one 50m line a single diver can effectively search the best part of 8,000 square metres. The only possible problem here is that of the line snagging on something on the seabed or catching in the weeds, hence the requirement for a reasonably clear bottom. Ideally, the line should be wound on a reel so that it can be let out carefully and there is less chance of being tangled up in great loops of line.

You can rely on compass headings to tell you when you have completed a lap, but it is not worth wasting the time to do this, apart from which you might easily miss something on the bottom while trying to check on your compass reading. Far better to take down a good clear bottom marker and lay it down in a prominent position. On completing the lap, pick it up and move out to the next swathe and so on.

Search techniques

Another advantage of this method of searching is that you can always note accurately the distance any find may be from your original anchor point. For this reason your anchor point must be firmly established and if necessary buoyed for future survey. In conditions of really outstanding visibility you can also establish a compass bearing of your find in relation to the anchor point. In less good conditions you can get a fairly accurate reading by holding the line tight and noting its direction as a compass heading. Do keep in mind that the bearing you are seeking is that of the object from the anchor point, ie the reciprocal to the compass reading you will get when looking from the object towards the anchor point. This is obtained by adding or subtracting 180 degrees from the reading. This is necessary, as when drawing up your charts you will know the position of your anchor point because you will have surveyed the buoy, marked it, and drawn it on your chart. From that point it will be easy to place the position of the find, but quite impossible the other way round. With combination of bearing and distance you will have a fairly accurate fix on your find. For this reason if no other, the sweep search is of great value and is to be recommended. Relatively little practice is required to carry it out.

TRAVERSE SWEEP SEARCH (p 110)

Another set of circumstances could arise, which might make either of the above search patterns difficult or inconvenient to use. This would happen if it were necessary to search an area which is long and narrow, for example, a river, canal, rock crevice, or sunken roadway. The use of square or sweep search here would mean constant movement of the start or anchor point as the range of movement is limited by the narrowness of the area to be searched.

In such circumstances, it is advisable to start by making a fixing, as before, in the centre of one end of the area to be searched. Its distance upstream, from the downstream end of the area, will be governed by the length of your line. Make your tape or rope fast to this and swim or allow yourself to drift downstream for the full extent of your line. Having reached the limit of the line, swim to one side of the search area; from here commence your search by swimming in a series of arcs from side to side of the search area. These will be relatively flat arcs at the commencement but will become increasingly curved as you wind the line in at the end of each traverse. The amount being wound in is dependent upon the length of your visibility. Once again there is the problem of fouling the bottom snags, but this cannot be avoided in this case except by swimming at a height which will clear them; whether this is possible is dependent upon the visibility in the area. The advantage of swimming upstream is that any disturbance the swimmer may make in the sand or silt on the seabed will be washed away behind him and will not jeopardise visibility still further.

Once again the position of any find can be related to a specific distance from the anchor point of the tape, which can be land surveyed, and again bearings either of the anchor point if visibility allows it or of the heading of a tightly held tape can be obtained. Such bearings should be treated with caution as obviously being all within a narrow arc of travel, they will be very close in their readings.

In all searches of the above nature, marker floats should be dropped wherever a find is located so that they can be surveyed from above water and should a large quantity of finds occur in a compact area, it is far better to employ a grid technique for the accurate recording of their position. The making of grids and their erection on site was described in Chapter 4, so little more need be said here other than about the need to relate the location of any grid to the search area. We are dealing in relatively small areas at a time, and, certainly if they are some distance offshore, conventional buoying and surveying from the shore involves considerable risk of large errors in positioning due to the smallness of the angles subtended. Probably the best plan is to relate the square or squares by underwater reference to the baseline, and lock it in that way.

The alignment of the grid should be parallel to the base line if at all possible and its corners should be measured and fixed by triangulation in relation to the baseline. Care must be taken to level it using the devices earlier described. I would stress that when a grid is put down, it must be very firmly fixed. It is infuriating if, half way through the recording of a complex set

Search techniques

of artifacts, some idiot comes charging across and catches the end of the grid with his foot or some other article of equipment, dislodging it completely. The earlier measurements are useless unless the grid can be repositioned exactly as it was before.

In fact, if a very long-term project is envisaged and grids are likely to become a fixture for some long time—this could apply on a wreck excavation—it might well pay to make up the grids with builders' steel scaffold poles. With these, verticals can be hammered very firmly into the bottom of the corners of the grid-to-be, and the horizontal members can then be levelled afterwards and fixed with scaffolding couplings.

CORERS (p 116)

While discussing work on the bottom, we may mention 'coring' as a method of locating items that are submerged under the mud. Coring is a development of 'probing'; in the latter steel rods are systematically pushed down into the seabed and note taken of the levels of any obstructions which are met. The extent of the obstruction will also indicate whether it is just a small article or much greater. It is also possible sometimes to detect differences between substances by the sound obtained by impinging the probe on to them. A metal or stone object will generally give forth a 'clink' even if some feet under the mud, whereas wood will not. On feeling an obstruction, continue probing outwards in one direction until the probe goes down further and is obviously beyond the extent of the obstruction. Mark that point and returning to the original probe hole continue probing outwards in all directions until the boundary limits of the obstruction have been defined by markers. This will give an outline pattern of whatever is below the mud at that point.

It is at this point that coring can be used to advantage. Here we use what is, in effect, a hollow probe in its simplest form. One end is provided with a relatively sharp edge which can be forced down and through any non-metallic or stone surface. On withdrawal, the inside of the hollow tube will contain a stratum by stratum picture of the surface which has been penetrated. Should you have pushed the corer through wood, then a plug of wood will appear at some level in the corer, telling

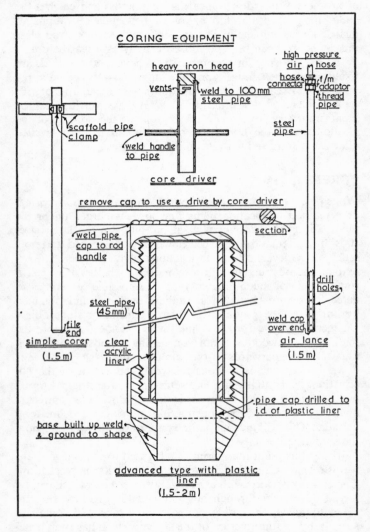

you the thickness, type, and condition of the wood below you. Considerable care must be experienced in the use of corers,

Search techniques

and indeed of probes, because either will destroy or damage any fragile articles they encounter. They should only be used in the last extreme. (Unless, of course, one is only trying to obtain a sedimentation sample of the seabed.)

A relatively simple corer can be made from steel conduit tube, of probably 15–20mm internal diameter. Some form of cross-bar type of handle should be incorporated, for considerable force can be required to get the corer down—and usually a great deal more to get it back again! More sophisticated corers have metal outer tubes but contain clear plastic liners. By using these, permanent records of the strata can be retained for, after coring, the liner may be removed and plugged. As it is transparent, its contents may be examined without removal. If the more simple corer is used, the cores can be pushed out on to a length of canvas, plastics sheet, or even metal foil and wrapped up for future reference. Provided they are stored flat and well supported, they may be kept for quite a long time. Full interpretation of a core is the task of a geologist and his services should always be at hand if coring is to take place.

On commercial corers, the outer metal casing is often left in the seabed and only the liner is reclaimed. By using this technique, coring can be done from a boat, with weighted corers plunging vertically down into the seabed under their own impetus. An automatic release device frees the corer and there is a flotation unit which returns the core to the surface, where it can easily be picked up. With these units coring can be carried out at great depths, well below those possible for a diver.

If much coring is to be done, it is as well to be prepared for the sometimes difficult task of getting the corer out without damage. A simple piece of apparatus can be constructed to assist in the removal of recalcitrant tubes. This consists of a piece of iron pipe, 20mm will do. Make a suitable connection at one end so that an air line from a bottle may be attached; a discarded single hose regulator which may be suspect for diving purposes will be ideal, using just the A-valve and hose.

AIR LANCE (p 116)

When a corer is difficult to withdraw, just push the iron pipe down alongside it with the air turned on. This becomes a very effective device known as an 'air lance'. The air escaping from the lower end will prevent it from becoming clogged and at the same time will aerate the surrounding mud etc, making it very loose. This will quickly loosen the surroundings of the corer enabling it to be easily withdrawn, but will not affect the contents.

The need for corers will probably be most likely to arise when searching for a wreck of a ship which is believed to be buried in the seabed and where a likely looking mound has been located. Careful probing will define any outline of substantial materials hidden under the mud. This need not cause much, if any, damage to the wreck. The next stage is to try to prove positively that it is the remains of a ship and not stone, iron, or even very hard clay, which can offer resistance to a probe that can easily be mistaken for wood of a very waterlogged nature.

Endeavour to establish from the outline, the main axis of the 'wreck' and at regular intervals insert the corer. As this will be passing down through what could be important structural members, the use of a corer needs to be as limited as possible, only sufficient to prove the presence of man-made objects below, and then stop! The length of the corer will normally be only 1–1½m; anything longer will not only prove difficult to remove but also to insert. The truth of Newton's Third Law (every action has an equal and opposite reaction) is really brought home to one under water. Usually the reaction of any endeavour such as probing or coring is a rapid movement of the diver, not of the object of his endeavours. One must be very heavily weighted if much coring is involved. This will help you to pull down the corer, for, unless the seabed is very soft, you will find this is the easiest way to start with. To facilitate this movement, it is as well to form a cross-bar type of handle at the top of the corer. The diver can then hang from this and so use his weight to penetrate the first half metre or so. After that it may be necessary to use a sledge hammer to really get it home (p 116).

Talking of weights, I have always been appalled by the amount of weight which some divers, particularly those who are novices, seem to find it necessary to wear. Even with only half suits on, I have seen divers carrying 7kg or more. This seems to me to reflect very bad breath control or semi-inflated life jackets! Whatever the reason, the effect is that once they are below about 3m, the divers are grossly overweight and when they reach the bottom they are for ever wallowing in it, and in the process stirring up silt and mud to the detriment of the visibility. It is far better for normal work that the weight is just sufficient to give a neutral buoyancy below 3m; this may mean being slightly buoyant above that level, but any diver worth his salt should be able to duck dive and swim down. It also has the benefit that when he returns to the surface possibly tired and cold, he is buoyant without the lifejacket.

However, there are times when a substantial degree of overweight is necessary on the bottom. We have mentioned this in connection with coring and it applies equally to other activities, including photography, where steadiness can be obtained by weight. Neutral buoyancy can be helpful here, but on balance I prefer to be slightly overweight so that I stay put on the bottom whether I am inhaling or exhaling.

To achieve this without being overweight on my return to the surface, I use old stockings. Yes, old stockings. Obtain a few pairs from mother, sister, girl friend, or wife and tie a knot in the instep of one, fill with sand and then tie another knot, to keep the sand in, about 6in from the top. This sausage of sand can then be slung around your waist and knotted in front of you. Put this on in addition to your normal 'neutral buoyant' weight belt. A full stocking can easily add 5kg or more of dead weight which will keep you firmly to the seabed. To come up at the end of the dive, merely undo the knotted ends and then the knot at the top of the stocking and pour out the sand—no more weight problems! It is surprising how easily the knots come undone on nylon stockings. Anyway, in an emergency, a prick with a diver's knife will split the stocking open and spill the sand. A lot better than jettisoning a normal weight belt costing—how much? It is worth a try.

So far we have discussed equipment well within the resources of any diving group, but mention must be made of more sophisticated equipment, even though some will never be available to Mr Average Diver.

METAL DETECTORS OR INDUCTION DETECTORS

Still on the bottom, a more elaborate piece of equipment which may be of value is the metal detector or induction detector. This is similar to the mine detectors used on land during the war and basically consists of coils of wire (well insulated) through which an electric current is passed, creating a magnetic field. The presence of a metal object will cause an alteration in the current due to the influence of its own magnetic field. This change is quite detectable and can be registered as an audio or visual signal. For use underwater, it has been found that the visual is more satisfactory than the audio which can lose its effectiveness because of other noises, such as exhaust bubbles or noisy valves, etc.

These instruments are not, at the time of writing, readily available and are somewhat limited in their application. They are primarily used as close-range locators and will register the presence of metals, if in sufficient quantities, at a distance of up to 1½m under the surface of the seabed. In the hands of an experienced operator it is claimed that different types of metal can be detected and identified, but it would have to be a very experienced operator.

Metal detectors are sold quite extensively in the US and Canada to amateur 'prospectors' to aid their hunt for gold, silver, and other precious metals. However, the best of these quote a maximum effectiveness in locating small items, such as coins, at a distance of less than 20–30cm, and that is on dry land! Underwater, it seems likely that the range would be no greater, if as great. However, this does not rule out their use in an area within which there is believed to be a substantial number of small artifacts such as coins, musket balls, metal jewellery, etc. They would be used in conjunction with a grid for accurate locating of any finds made. The 'hot spots' having been pinpointed with the instrument and marked, a grid would be placed over the area. As each spot is investigated, any object that may be found can be immediately and accurately located in relation to the grid. In larger areas where a grid of tapes is being used, as described on p 177, the instrument may be operated over the grid.

Probably their greatest disadvantage is that they will give erroneous readings due to the presence of large masses of high magnetic influences such as volcanic rock, ironstone, and of course, in rivers etc, the inevitable debris which only man in his greatness could fabricate and use to desecrate, items like car bodies, bicycles, bedsteads, and tin cans! In harbours they change slightly to hawsers, anchors, chains, and—tin cans. Work is continuing on this type of equipment and it is more than probable that as time goes by more accurate instruments will be forthcoming and will be a boon to underwater archaeologists.

MAGNETOMETER

A similar device, but one which works over far greater ranges and can therefore be operated from a boat, is the magnetometer. As the name suggests, it works by detecting variations in the earth's magnetic field which could be caused by ferrous objects. Unfortunately, as yet, it has not proved to be an entirely reliable device nor is it capable of analysing or quantitatively defining what it registers. For instance, a large ferrous anomaly some distance away can give the same degree of reading as a very much smaller one which is somewhat nearer. Again natural anomalies such as volcanic rock, ironstone, etc will give readings which can confuse the searcher. However, it is one of the less expensive of the scientific instruments that underwater searchers can bring into use and as such, merits consideration. The fact that it can detect when visual search is impossible is of the greatest importance and constant developments are taking place in the design of these units, all aimed at a greater degree of definition of the find, greater screening from distracting influences and greater reliability. Those who are likely actually to use these devices will have to go far more deeply into the subject than we shall here and there are far more exhaustive publications available from which they will be able to obtain their information (see bibliography).

Probably the most common—if one can use the expression —type available today is the 'proton' magnetometer which is reasonably well proven, but developments with other sensors are currently being made which should improve the device still more. A typical instrument consists of a transistorised electronic unit and a sensing head. Usually relatively light in weight and self-contained, it is entirely portable. Power is supplied from a battery cell, for lightness probably a nickel cadmium one. The sensing head can consist of a bottle containing distilled water (or any hydrogen-rich liquid) around which is wound a coil of wire. This water is the source of the hydrogen atoms of which the protons are the nuclei. The sensing head and recorder package are connected by a shielded and waterproof cable which should be fairly long, probably 30m at least, so that if necessary they can be widely separated. In fact, the best results can sometimes only be obtained if the actual sensor head is towed behind the search vessel on an inflatable rubber dinghy which has been stripped of all metal fittings.

In general principle, the instrument works like this. The water in the sensing head contains a proportion of protons that are precessing, as does a gyroscope, but without phase coherence and in random magnetic orientation. A strong field (100 oersteds) at right angles to the earth's field is induced by the coil, causing the protons to align themselves magnetically. When the induced field is switched off the protons will precess in phase. The same coil can then be used to detect the precession signal/field strength.

In surveying the *Amsterdam* in 1970, it was possible to detect large masses of iron in both bow and stern. Mr Jeremy Green, who undertook the survey, was able to calculate the approximate weight and depth of these masses.

The combined use of the metal detector (capable of ferrous and non-ferrous detection) and the magnetometer (capable of ferrous detection only) in the same search very quickly establishes whether a mass is ferrous or non-ferrous. If the former, a reading will be obtained on both instruments, if the latter it will only be picked up on the metal detector. More detailed information on these very scientific techniques may be obtained from publications listed in the bibliography.

The comparatively recent development of prospecting for oil beneath the seas of the world has brought with it rapid growth in many techniques aimed at making the searches quicker, more accurate, and less costly, although what may be

Search techniques

'less costly' to an oil company is probably still miles out of the reach of you and me!

Many of the developments have also derived from naval detection devices based on the sonar principle. This—to give a greatly simplified explanation—requires the sending out of a signal, the receiving of its echo and the measurement of the time interval. Divide this by two and the speed of the sound under water and you have the distance of the object from the transmitter or transponder, as it is called. The fairly commonplace 'echo sounder' for telling how much water there is under the keel uses this principle. More precise instruments can detect much smaller and less well defined objects than the bottom of the sea, for instance fish shoals, and most large trawlers are equipped with this gear.

ECHO SOUNDERS

These can certainly be used in the search for underwater remains, but it is rather a hit-or-miss technique, for they project their sound pulses more or less directly downward in a fairly narrow cone. Therefore coverage of a large area of seabed requires considerable skill in navigation to ensure that the whole area is completely covered.

Some experience and indeed skill is also required to interpret the signal received on the instrument. All that can be seen is a narrow band of light on the depth scale. The sharpness of the band will indicate the type of bottom, for instance, a level, hard bottom will give a very clear-cut narrow line, whereas a soft muddy bottom will produce a broad trace, probably blurred right across. If there is a narrow layer of mud overlying hard rock, the deeper edge of the trace may well be defined as a hard line, but this depends entirely upon the depth of mud and strength of the instrument. A bottom covered with a mass of large boulders would again show as a wide trace with a clear hard line at the top and a lighter, 'ghost' line below it; this assumes a fair degree of coverage by the boulders. Occasional large rocks standing up from the bottom will be indicated by the depth line on the dial moving up and down as the height of the bottom varies. A wreck standing out of the seabed will show a similar trace. Its presence would initially be

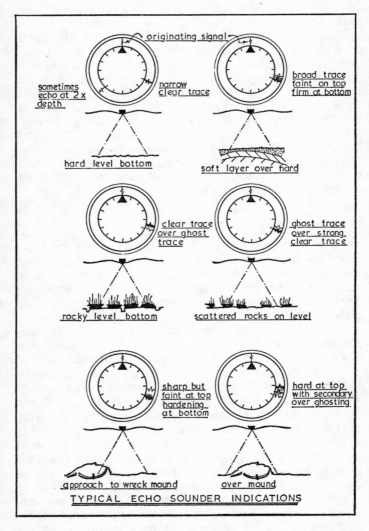

indicated by a trace of the bottom in the form appropriate to its consistency, which would gradually become a broader line

Search techniques

as the fringe of the echo sounder beam picked up the wreck, with ghosting at a shallower depth reading. This in turn would harden into a firm line at the shallower depth as the transponder passed directly over the wreck and then would gradually fade away again. These indications are sketched in the fig opposite.

The position of the transponder is important because it must be free from extraneous noises or echo producers. Turbulence and bubbles are sources of such interference, so it is important to site the transponder where these effects are least. Generally on a large craft, the transponder is let into the bottom of the boat and care should be taken to keep it free from weed growth and it should not be painted over. It is possible to obtain portable units which can be clamped to the side of a small craft, but be sure that they aim vertically downwards. It is useless to align the unit vertically, and then motor along with all weight aft or on one side of the boat. The resultant change in fore-and-aft levels or list to one side would cause the echo sounder to over-read, as the sound waves would not be going vertically down to the bottom. Also, with portable units, do see that they are well waterproofed, as the 'works' are easily upset by damp.

SIDE SCAN SONAR

This uses the same principle as an echo sounder, but the beam of sound is directed obliquely towards the bottom. The range of a modern unit can be in excess of 300m and the transponder is generally towed in what is known as a 'fish'. It is therefore possible to project the sound waves to both sides simultaneously and so in one pass record the bottom profile to either side of your craft.

A height to *horizontal* range ratio of 1:2.5 to 1:10 is recommended to obtain the best results in most applications. In effect, the more horizontal the beam the better the picture obtained. This is because the beam is not being reflected by the mass of the bottom, but by the irregularities upon it. For instance, large mounds, rocks, or solid wrecks will reflect the sound waves back, but because sound waves, like light waves, move only in straight lines, they cannot penetrate or bend round the obstruction and therefore a 'shadow' is left behind it. On the trace, this appears as a white area which outlines the shape of the obstruction fairly accurately. A combination of this white outline and the comparatively heavy dark readings obtained from the obstruction, indicates quite clearly that there is an anomaly. A number of traces across this obstruction can show its section, height, and length and much can be deduced from this.

The definition of the bottom pattern is affected by the speed of the towing craft, the height above bottom of the transponder, the distance from the object and the speed of the paper through the recorder upon which the trace is being made. It is, of course, necessary to relate the speed of the towing craft to that of the paper to avoid producing distorted patterns. In deep water, which we are not really concerned with here, it is general to tow the 'fish' at a height of approximately 60–70m above the bottom.

From the foregoing, it will be recognised that considerable technical knowledge is required to use this equipment and to interpret the records accurately; this, coupled with the cost of the equipment puts it beyond the reach of the amateur operator. Practically every month sees new developments in the design of this equipment and new publications are constantly appearing which add to the knowledge in this field, and the reader is advised to turn to these if his interest is more than casual (see bibliography).

SUB-BOTTOM PROFILE TRANSDUCER OR 'PINGER'

Frequently it is necessary to know what the bottom is like below the actual surface face; in the search for oil—or in the case of a wreck that has been completely covered—this is paramount.

The *Mary Rose*, in the Solent, is such a case. As far as can be ascertained by diving, none of her remains are visible above the surface of the mud bottom; something is required to 'see' through the mud. A side-scan sonar may well define an anomaly in the form of a mound of interesting shape. The sub-bottom 'pinger' will penetrate down through that mound and will indicate the depth of object which has caused the mound

Search techniques

to form; it will also indicate its outline so that the combination of both instruments can provide a formidable amount of information—but it will all need expert interpretation.

One limitation of an SB-pinger is that it scans a relatively narrow track across the bottom and therefore would require many tracks, accurately plotted, to ensure adequate coverage, a slow and costly process. Using the side scan first can cut this time factor down to more reasonable proportions.

The depth of penetration into the bottom is controlled by various factors; it can be as much as 150m, but is usually somewhat less and is dependent upon the acoustical properties of the sub-bottom and the choice of the correct frequency, for the sound wave. The higher the frequency, the less the penetration. The low-frequency capability is only limited by the size of the transducer. Depth of water, salinity, and surface conditions all play a part in the accuracy of the results.

To obtain the higher-energy and lower-frequency sound source that is necessary, we have to depart from the conventional 'pinger' of a depth echo sounder and use a 'sparker', 'boomer', or air/gas gun.

The sparker system consists of a charging circuit with a suitable power supply, an energy storing device, high current switch, transmission line, and an underwater spark transducer. Very large quantities of energy can be stored and released at rapid intervals in the form of acoustic pulses with pressures up to 130 decibels.

A boomer requires a high-voltage supply source. Stored energy is discharged into a transducer coil where it repels what is known as an eddy current plate. The plate is repelled from the coil at a tremendous rate and it is the impact of this plate against the water which creates the sound wave, rather like hitting the surface of the water with a flat board or an oar, which we all know makes a loud clap.

Gas chambers depend upon the ignition of a gas mixture in a chamber submerged in the water, the resultant explosion creating the sound wave. The air chamber unit is safer and cleaner but requires a compressor to compress air in an underwater chamber. The air is suddenly released by a solenoid valve and the escape of this air produces a low-frequency sound wave of very large amplitude. The only limitation with this system is that of the speed of recompressing the chamber and this in turn depends upon the air supply from the compressor. However, the system is increasing in popularity in the oil prospecting world and anyone who has something over £1,000 for the 'fish' alone, can obtain a unit. Before spending that amount of money it would be advisable to refer to some of the more advanced publications concerned with this type of equipment, which are listed in the bibliography.

In this chapter I have tried, if only briefly, to describe the various techniques that may be used to search an area. All require considerable practice and it is as well to attempt trial runs in the open sea without any specific objective to familiarise the team with some of the problems, before actually going on to a site. Who knows, on a dummy run you may well find something of importance.

7
THE SIGNS WE SEEK

THE MASTERY of techniques will be of little value if we do not know what we are seeking. It is all very well to say we are looking for a wreck, harbour, or what-have-you, but as pointed out above, these will not be lying on the seabed in all their glory with little arrows pointing to them! Detection of wrecks involves the noticing of little things which can give a clue, the diligent following up of this clue, and the ultimate reward of finding the quarry.

The sea, its inhabitants, and vegetation all conspire to hide from prying eyes all that has once been offered to it. In a matter of weeks, weed will grow in profusion, in months calcareous deposits will begin to appear, in years the combination of both will camouflage anything, softening the outline of a man-made object into an indistinct 'no shape' over which the uninitiated will swim without a second glance.

Consider how wrecks, our most likely target, come to be. If one ignores those caused by act of war, where shell or bomb has burst a vessel open and caused it to sink in deep water, the majority have come to grief through hazard of storm and rocks. Possibly through an error in navigation under the stress of terrible weather conditions, a ship strikes a rock. She is holed and starts to sink; her anchors are probably let go to try to hold her on the rock, her heavy top hamper and cannon are possibly jettisoned to reduce her weight. Despite this she may still slide back into deep water, eventually reaching the bottom many hundreds of metres from her original point of impact.

On another occasion, a ship strikes a glancing blow on a submerged rock, she is holed and starts to sink, again her skipper tries to lighten her load but all to no avail, and again her resting place on the bottom is far from her point of impact. In the seas off the coast of America, the coral reefs were the cause of many a disaster. Rising sharply out of the depths to within only a couple of metres of the surface, they trapped many an unwary mariner. Often going full speed, a ship would strike, her bottom would be ripped open, and the contents strewn over the seabed as she careered on. Her path would be marked by crushed coral, where she had cut a groove over the reef before sinking into the deeper water on the other side. These marks can still be seen hundreds of years later and have led many a hunter to his quarry in those waters.

We have cited some of the many circumstances that can be associated with the wreck of a ship. What is there in common? Firstly, that the final resting place of the wreck can be far from its point of impact with the rock which caused the tragedy. Secondly, that her downward path to the bottom will be marked by a trail of debris which has been either deliberately or accidentally jettisoned from her. Thirdly, her anchor will possibly be imbedded in the reef, rock, or seabed with the shanks pointing along her final line of travel. This latter point is a very significant clue, for even if a ship has been driven on to rocks from being at anchor (the *Santa Maria de la Rosa*, was such a case) they will be imbedded in the bottom, having been dragged along until the hull itself stopped moving, and again the shanks will indicate the line of travel.

In Sicily, we found the wreck of the comparatively modern *Llanishen* (1875) with her anchors and chains stretched out in front of her bows. She had struck the reef and her skipper had tried to hold her from slipping back into the deep water. Once more the chains and anchor shanks indicated the direction of her last fateful glide into eternity.

It therefore follows that any article found on the bottom of the sea can be a clue to a much greater find. It is of paramount importance that it is not lifted, or even moved, until its exact location, bearing, and relation to its surroundings has been most carefully noted. In the Mediterranean, many a wreck of ancient times has sunk into the soft bottom and its cargo mound

The signs we seek

has gradually been covered over. Perhaps the neck of one amphora still protrudes above the sand. Along comes a diver with shouts of glee, digs it out (probably breaking it in the process), and rushes to the surface. The sand gently sifts back into the depression, the slight surge in the water even at that depth, smooths over his scraping and within a very short time, all is serene. And so, forever, the wreck will remain undisturbed because it will never be found again—our diver friend has removed the last sign of its presence.

It is absolutely essential, and this cannot be stressed too strongly, that before anything, anything at all, found on the bottom is moved, the finder should mark it, 'fix' it, and commence a search from or around it in an endeavour to locate other clues.

Common sense, a knowledge of prevailing winds and currents, and the study of likely points of impact within a reasonable distance of the find, will all help to orientate it and lead to the location of other clues. What are the most likely clues to be found? Obviously those that withstand the action of water, and insect or animal attack. Again, those that are relatively large are most easily seen. Amongst these are the anchors already mentioned and which in themselves will tell much about the age and size of the ship from which they originated. Because of their very distinctive shape, they are less easily camouflaged by nature and are consequently more easily seen by the diver.

Cannon probably come next in ease of location, but do not be too confident about this. Because they are extremely heavy they will sink quickly into soft bottom and possibly only a small portion will still be showing. This will be covered in time by weed, molluscs, and concretion and will only be detected by taking great care. Even on a rocky bottom, where generally fair amounts of weed grow, they will be very difficult to recognise. At SNAP, there is a marked-out search square, probably about 30m x 30m, in which there are a number of large cannon, concealed in weed and wedged between rocks. Few divers give an accurate count of the actual number there until several attempts have been made. When visibility is down to 1–2m at best, even large cannon are easily missed.

The clues to look for with cannon, are the straight lines of the barrel, or tube. Anything that is remotely straight amongst the general roundness of a boulder-strewn bottom warrants close examination. It is surprising how often the hole in the muzzle will stay clear, sometimes acting as a home for a moray eel! The round darkness of this opening will easily catch the eye of the diligent searcher. The round ball of the pommellion at the breech end of the cannon can also be spotted. A lightweight geologist's hammer is a useful tool; a gentle tap on any suspect piece on the bottom will produce a ringing or clinking sound. That from a metal object will be significantly different from that of a rock. A hollow article will sound different again. Only do be careful how hard you hit the object—it would be a tragedy if it were an amphora or valuable antique pot! I have used this technique many times in searches for amphoras, and with great success.

Cannonballs are often found in great profusion in coastal areas—either fired from a shore-based battery or part of a jettisoned cargo. A trail of these may well lead to a wreck. Not the easiest of articles to recognise, they do have the benefit of generally being fairly free from weed growth and reasonably round. The smooth surface of a new ball may well be badly corroded and the outer surface turned into crystallised iron oxide with large cracks penetrating deeply into the surface, so do not be put off by that.

A point to be remembered about any metal objects is that they have different electrode potentials, dependent upon the type of metal. When dissimilar metals are immersed in sea (salt) water which acts as an electrolyte, they become, in effect, a galvanic battery. A flow of ions (ie, electrically charged atoms) will take place between opposite poles, and when this happens under the sea small particles of sand, sediment, and calcareous matter will be carried along. Eventually these will be built up on the surface of one of the metal objects and will completely encase it in a crust of almost cement-like material. Fortunately, the original shape of the object will not be distorted too much and should be recognisable. In waters where coral is present, this is not so, for coral will grow upon the calcareous covering and can camouflage it completely.

This accumulation of deposits sometimes forms a hollow mould around the original metal article, which has long since been converted into oxides. If carefully opened, this outer covering can be used to form plaster replicas of the original content.

Where two metals of significantly different electrode potentials are close together, frequently the one with the lower (ie, less negative) potential will escape the major effect of the electrolytic action. Many readers who may have boats will know of this as 'cathodic protection': a piece of zinc is placed in close proximity to, say, the brass propeller of the boat to prevent the propeller from being attacked. This effect can account for the relatively undamaged piece of metal sometimes found in contact with another of greater potential, eg pewter which may have been in contact with iron.

On seabeds covered with boulders, search carefully deep down among them, looking into all the cracks and crannies and, indeed, rolling stones to one side, so that you may examine the bed underneath them. Small artifacts of heavy nature, such as coins, jewellery, metal fittings, will tend to work their way to the lowest points. Once there, the currents and surges of the sea will not disturb them, but will tend to wash sand away thus leaving them fairly well visible. Quite large boulders will move over a period of years and it is quite possible that one may have rolled into a depression in which some artifact is lying, so concealing it from view.

On a naturally boulder-strewn bottom and even more so on a sandy one, keep a sharp look out for an 'unnatural' pile of stones. Many of the ancient ships used rock for ballast, and in some cases those which were nearest to the ship's sides were cut to shape. In all cases the stones would be graduated in size from large blocks to relatively small pieces that would have been wedged between the larger ones to prevent movement. Unless the bottom had been torn from the hull, these stones would still be more or less in their original position relative to the hull shape, but would obviously spread wider as the walls of the hull rotted away. However, such a pile is quite distinctive and will stand out even among natural boulders, therefore any compacted pile of stones should be investigated. A geologist on your team will be of great assistance here.

The likelihood of finding traces of the ship's timbers is fairly slight. Most wood structures that have not become buried by the mud and sand of the bottom will over a period be eaten away by the mollusc—*Teredo Navalis*—which is generally referred to as a 'shipworm'. However, in waters which do not encourage this mollusc, such as those in Stockholm harbour where the *Wasa* lay, the walls of the ship may still be relatively intact. More generally, it is those that have sunk into sand and which have been quickly covered by silting, which retain any substantial amount of timber. Upon excavation, such timbers may come to light after the removal of ballast stones, etc.

Steel-hull vessels of more recent times will retain their form to a large extent, but these are not likely to be of sufficient archaeological interest to warrant discussion here. However, in passing, I would advise great caution if driving within such a hull, particularly if it has been sunk by wartime action. Torn plates, under the action of the sea, can become deadly sharp and in the darkness of the hull it is all too easy to swim into such an edge with, possibly, most unfortunate consequences. Should you dive within such a vessel do ensure that you have adequate air, torches, and possibly even guide lines. It is easy to lose one's orientation in the excitement of exploring the complexities of the ship's construction, so it is well to ensure you can quickly find your way back to your entrance point. There is also danger from the sudden collapse of metalwork due to a diver accidentally knocking against a badly corroded support and snapping it through. This is one of the worst hazards of salvage divers.

We have covered most of the obvious clues which can indicate the presence of a wreck, but there are many other but smaller indications that may be found. Summarising, it can be said that anything that is foreign to its surroundings, should be investigated and treated not as a 'one-off' find or item but merely as a piece of a jig-saw. From that one find, spread your search until you find other clues or else establish that it is in fact just a casual find.

Other more subtle clues that can lead to a wreck location are perhaps more likely found in a smooth sandy bottom or mud bottom. Here, the hull may well have sunk out of sight, but in doing so it will have left some mark on the surface of the seabed. This could equally well be a shallow depression left by the eddy of the tide or current over the obstruction, usually

The signs we seek

found down current from the cause, or it could be a mound that has built up over the wreck. The latter is perhaps more likely; many such have been found in the Mediterranean covering Roman and older wrecks. Should such a sign catch your eye, then do investigate it. Again on a sandy bottom, the indications of a straight or gently curving line, looking almost like a narrow crack in the sand, could prove to be a piece of deck beam, rib or other part of a ship.

In looking for remains of harbours, we seek many of the same clues that have been already mentioned. Anchors will possibly indicate an anchorage, where ships have waited prior to entering a harbour. In other cases, particularly the harbours of Phoenicians, there were no outer sea walls and only quays against which a ship moored for unloading. Dependence upon the land mass of an island or peninsula to provide shelter from prevailing winds was common, and vessels anchored in the lee waiting their turn at the quay.

The anchors found in such places could range from the type with which we are fairly familiar, such as those used on seventeenth- and eighteenth-century vessels in this country, to a very different type, the old stone anchor. These, in their simplest form, were pieces of stone through which a hole had been made to take the bight of the anchor rope. It should be noted that all stones with single holes were not necessarily anchors; they may have been line sinkers used to weight down the anchor line. The alignment or grouping of a series of stones could be very important and should be accurately recorded. Others, rather more triangular in shape, had one hole at the top or narrow end and others in the wider base end, through which wooden stakes would have been driven to act as the equivalent of the modern fluke (see figure, p 264). Such anchors have been well catalogued over a wide date span, so reference to one of the books (eg by Miss Honor Frost) listed in the bibliography can help in dating a site.

Wherever craft were anchored, rubbish was jettisoned over the side. Today it is beer cans and 'Coke' bottles, then it was amphoras and broken pots. All of which can help to piece together knowledge of trade routes and usage of the anchorage. But they do not necessarily indicate the location of the port itself.

The search for amphoras themselves can be quite teasing, for they too very quickly receive a covering of encrustation and when partially covered by mud or sand are indistinguishable from round stones. So we revert to the geologist's hammer, mentioned earlier, and tap lightly any suspect object. If made of earthenware, it will emit a very satisfying hollow note. Many an amphora has been missed by a diver swimming too quickly over it. One very successful amphora-hunter I knew, would always take his flippers off when he was in the search area. Without them it was impossible for him to swim fast and he therefore had far more time for his search! In shallow-water work, we often dive from the boat without fins, for another benefit is far less disturbance of silt and sediment which would destroy visibility.

The identification of actual harbour constructions themselves is not so much a case of looking for clues to help locate them, for these can be quite immense and indications can be seen from the surface, indeed from aerial photographs and sometimes from the top of high hills. The problem here is to assure oneself that one is, in fact, looking at a man-made structure and not natural rock formation, or if it is the latter, that man has been at work on it to adapt it to his own requirements.

In many cases, where seaward defences were made, generally these utilised existing reef formations which were adapted by the cutting of steps on the seaward side to break the effect of the waves and the stone blocks so quarried were often built into walls on the top or level quay areas on the lee side. However, the action of time and tide on stone affects both man-cut and natural un-cut rock formations equally and they become almost impossible to tell apart. In many areas the indigenous rock is formed into strata so regular in size and shape that many a diver has found what he first thought to be a wall of squared rocks, only to realise eventually that it was a natural formation.

One of the first things to check is whether the stone forming the blocks is indigenous or whether it is different and therefore brought to that place. Sandstone and limestone were very popular building stones, for they were easily cut and shaped. Frequently they were brought from quite distinct quarries for use in building and were laid on to the levelled bedrock, which

The signs we seek

in many cases may have been volcanic, so the difference between the two is easily detected. Notice also the angle of stratification of the rock you are examining and that on the neighbouring land; if it is very similar be careful and double check that it is not natural. If strikingly different, it may be a good sign, but local earthquake action could have caused a fault which would have made the difference, so even that is not infallible. Once again the knowledge of the geologist will help here, and if necessary break off rock samples and take them to a geologist for identification.

Check carefully the apparent joints between the blocks: are they fairly regular in spacing, ie are the blocks similar in size? Are all angles square and edges reasonably straight? The Phoenicians were expert masons and cut all their stone blocks so that they could build without even using mortar to joint the blocks together. Look for the tool marks on the surface of the blocks; these are not easily seen but can sometimes provide conclusive evidence. Frequently large blocks were keyed together with lead 'plugs' and the hewn, tapered dovetail slots can be easily recognised. Note the actual size of the blocks. For substantial buildings, roads, and harbour works, blocks measuring 2m x 1m x 1m were not uncommon, and all would be well shaped. Look for indications of rectilinear patterns on the bottom such as would be formed by the remains of walls and parallel lines that could have been the edges of the roads or paths.

On smooth sandy bottoms, thin lines of weed can sometimes appear, which are growing in the cracks in paving stones some few centimetres beneath the sand and are evidence of a hidden structure. On heavily weeded and muddy bottom, one often finds completely vertical walls which rise as much as 3m from the bottom. These 'walls' can have sharp right-angle bends, may even end in a perfect box shape, and are calculated to mislead the majority of divers into believing they are, in fact, walls of man-made construction. But regrettably, more often than not they are natural weed banks which currents have cut into perfect geometrical shapes. On one occasion, some of my team, not convinced by me, casually said they were off for a 'recce' dive. They reappeared nearly an hour later, very tired and despondent—they had laboriously cut their way clean through one of these banks, coming out the far side. They were convinced by that time that I had not been kidding them! A strange feature of these banks is that they can alter remarkably over comparatively short periods of time. Certainly from one season to the next, and even occasionally after a particularly violent storm, they will change their direction and even move several metres across the seabed, yet still retain their angular shapes.

The tendency when swimming near these looking for 'finds', is to swim along the bottoms, which are frequently undercut. It is possible to find sherds in such places, but do not ignore the banks themselves. In 1967 at Nora, one diver noticed a round smooth object sticking out of the bank, halfway up and was casually scratching at it with his knife, when he noticed the strange behaviour of his buddy, who was on the other side of the object. His friend's eyes were practically out on stalks and he was gesticulating violently. The other side of the 'round object' had the perfect features of a Roman head—and it was one! A beautiful, unmarked, life-size terracotta head of about first century BC.

It should not be thought that sea walls, large buildings, roads, and quays were all built with these large blocks we have mentioned. The Romans built in a very different manner to their predecessors the Phoenicians. They preferred the less laborious method of using small stones, rocks, broken tiles, bricks, and even seashells with which to create their structures, depending upon the intensely hard mortar, which they perfected, to hold and bind the whole together. Frequently, one can clearly detect the structures of the Romans overlying those of the Phoenicians by these differences in construction. However, the Romans were not above using the stone blocks, hewn by the earlier masons, in their own constructions.

It is not uncommon to find the remains of a Roman mole or jetty lying as a mass of quite small rubble on the seabed. Generally this will have retained the rough outline of the original structure, and, if on a sandy bottom, this will be clearly seen. Once again the presence of many sherds, tiles and bricks amongst the rubble will be a further indication of the Roman handiwork.

In harbour searching and surveying it pays to use common

The signs we seek

sense or seamanship when commencing the search. The prevailing winds have changed very little in their direction over the centuries; the Sirocco still blows hot and moist across the Mediterranean from North Africa, the Levanter still blows as strongly from the east as it did twenty centuries ago, consequently a land mass which offered shelter then, will do so today. Small craft which used its shelter then will do so today; so, you will find that the local fishermen still anchor in almost the same spot that their ancestors did all that time ago.

Consider then, where you would have put a jetty to provide even more shelter or where it may have been safe to bring shallow craft close in-shore so that they could be unloaded. Search those areas and you may be rewarded. Note the currents and prevailing winds again, this time in relation to headlands that may be near the anchorage. Ships in those days could not beat against the winds, they ran before or on a broad reach across them. Some of the ships, trying to make the lee of that island or headland, might have been caught in the current you have noted and that, combined with the on-shore wind they were trying to obtain shelter from, would have driven them on to the rocks. Put yourself in their place and imagine where they may have ended up, dive there and you may again be rewarded with a wreck or amphoras.

Common sense is one of the greatest assets that a diver can possess! Imagination is a close second.

8

LOCATING AND PINPOINTING THE SITE

So FAR we have examined the situations leading up to the finding of a site, be it wreck or harbour. We have established that preparatory work is a prime ingredient of success and now, having located a site, we wish to set about recording it.

Assume for the moment that we have found a wreck, or an amphora, anchor, or what-have-you. As already stressed, it is important to leave it there until we have ensured that we can not only find it again but also anything in the area that may have been associated with it. Research may have guided us to the approximate area, but it could never have pointed us precisely on to a comparatively minute article, in hundreds of square metres of open sea. Now, we must be absolutely certain that we can find that exact spot again.

If possible, the first thing we will do will be to buoy it and so have a marker floating on the surface. But we may not have a buoy with us and even if we do, we must note its position so that, should it be lost, we can find our way back to that spot on our next dive.

So we look for landmarks on the shore that will help us. Excellent, right opposite us on the beach is a house and by sheer good fortune, just a little further along there is a tall tree. Fine, so we know that if we are opposite the house and the tree, we shall be over our spot—or shall we? Many divers will say yes, but they will be extremely lucky if they are. We may be within, perhaps 50m of the spot, but if visibility is down to even 10m, we will not be able to see our find. No, we have only half done

Locating and pinpointing the site

the job; the right idea but not fully developed. We must be more precise and use 'transits'.

TRANSITS (p 142)

Transits or leadlines are used in navigating craft in difficult waters. The helmsman can see two markers. There is perhaps more than a kilometre between them and he knows that by keeping them directly in line, he is keeping to a very straight course. For in fact, counting his own eyes, he is lining up three points and the slightest deviation from straight will show.

Much the same method is used when locating a site, but instead of looking at a large object such as a house, select one feature on it, a chimney or window edge, and line that up with an equally specific point either in front of or behind it. The second point might be a mountain peak, telegraph pole, or tree, it doesn't matter so long as it is distinctive. The observer's position is on an extension of a straight line joining those two points and will be accurately known to within a metre or so. Now look at least thirty degrees to one side or the other and pick up two more in-line points, noting them equally carefully. It pays to jot down the references on a writing pad to refer to on the next dive, for frequently one will not be able to remember just which points were chosen, especially when they all look so similar from water level.

When next coming to the site, steer your boat along one transit line with the helmsman ensuring the marks are exactly in line, while another crewman watches and waits until the second set line up exactly. Carefully done, this will put you right over the correct spot with no error at all. This, after all, is how your fisherman friend can always take you exactly over the wreck or fishing bank each time you go out with him. He knows his coastline and his transits.

PHOTOGRAPHIC TRANSITS

Given a camera, permanent records of the transit marks can be taken in a photograph of each set. Sometimes both sets can be registered on one plate, and there may be several sets

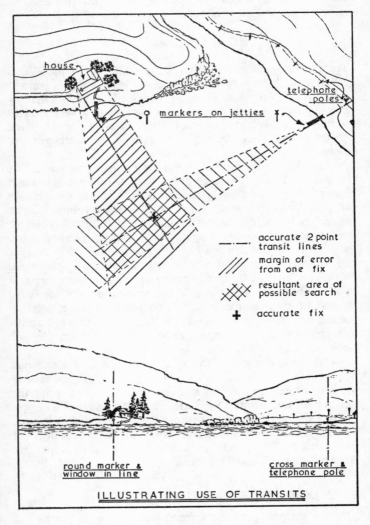

appearing to add to those originally chosen, so great accuracy can be obtained, by a 'multiple fix'.

Locating and pinpointing the site

COMPASS BEARINGS (p 144)

Transit lines can be obtained by taking compass readings from your position on to prominent landmarks. A series of three readings on to points at least thirty degrees apart will give moderately accurate fixes, depending upon the capabilities of the observer, the size of scale and accuracy of the compass, and the calmness of the sea! But for our purposes, I do not believe even the best will be sufficiently accurate to pinpoint a really fine plot.

Compass bearings are useful for navigating a boat, where 10–50m inaccuracy is unimportant when making a passage down a coast, but by no means good enough for charting purposes. However, a compass may be the only instrument available, in which case it will have to serve, extra pains being taken to obtain precise readings. But do remember that the bearings are from site to shore. When transferring them to a chart other than a naval chart, it will be necessary to draw in the reciprocal bearings, ie the reading you have taken plus or minus 180 degrees, from the landmark shown on the chart towards the site your endeavour is to fix. If you are using a naval chart on which a compass rose appears, you may apply your compass bearings without correction. In such cases, parallel rulers are best used. Place them so that one edge passes through the centre of the compass rose and the bearing obtained by the compass reading, extend the rulers until the second ruler passes through the landmark and then draw a line from that landmark towards the site area. The three lines drawn from the three landmarks will form a 'cocked hat' over the area of the site. It goes without saying that all landmarks chosen for reference points should be permanent and shown on the chart, also bearings taken with a compass are purely magnetic bearings and would have to be corrected for annual variation to provide the True bearings to be applied to your chart.

SEXTANT

This is the accepted seaman's instrument for establishing a position, for with the sextant it is possible to measure with

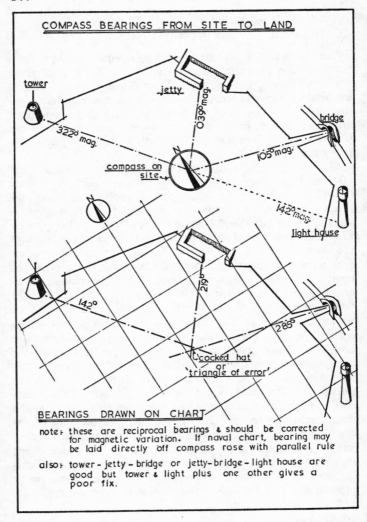

very great accuracy an angle, either in the vertical or horizontal plane, between two points. The navigator will 'shoot the sun', ie measure the angle between the sun and the horizon to deter-

Locating and pinpointing the site

mine his longitude and latitude—we can measure the angle at our position subtended by two on-shore fixed points. A proper navigator's sextant is a very delicate instrument which can offer extremely precise readings in the hands of an expert. It is also very costly and not really the most suitable piece of equipment to take out in a small open boat. However, recently a sextant made in plastics material has been put on the market in UK by Ebbco Ltd and can be obtained from most ship's chandlers for under £10. It is accurate down to one minute of angle (a sixtieth of a degree) which is quite adequate for our purposes. Supplied in a tough case, it is sufficiently robust to be carried on survey work, but even so, it does pay to take care in looking after what is, after all, a delicate measuring device. Complete instructions are provided with the instrument. These explain very simply how it should be used, but practice is advisable!

The following, in conjunction with p 146, will explain the functions of various parts of the instrument and how to use it.

A sextant consists of a 'plate' to which is fixed a movable bar called the 'index bar'. One edge of the plate is curved and is called the 'limb'; this curve is one-sixth of a full circle and is divided into graduations representing degrees from 0 to 110, although each, in fact, measures only half a degree. However, the system of reflection of light uses the same beam twice, which doubles the angle and so each graduation effectively reflects a whole degree of a circle.

Halfway up the plate is set a telescope which has its line of sight exactly centred on a mirror ('horizon mirror') which is fixed to the face of the plate and at right angles to it. The horizon mirror is only half of a square of glass, the other half being unsilvered, so that the observer, looking through the telescope, sees through the clear glass to a distant landmark and at the same time sees a reflection in the mirrored half.

Another mirror, the 'index mirror', is attached to the index bar directly over the pivot point and is again at right angles to the plate. This mirror moves with the index bar. The purpose of the index mirror is to reflect an image on to the horizon mirror and so back through the telescope to the eye of the viewer.

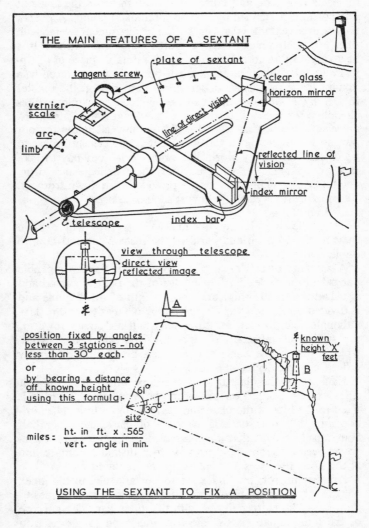

At the bottom end of the index bar is a minutes scale graduated from 0 to 60; this scale reads as a vernier scale in

Locating and pinpointing the site

conjunction with the degrees scale on the limb and from it, minutes of degrees may be read. Thus any movement of the index mirror and therefore the index bar will automatically register against the scale on the limb.

To use the sextant for positional readings, hold the instrument with the plates horizontal, index bar uppermost. Disengage the worm screw of the vernier scale so that the index bar is free to move. Sight through the telescope and through the clear half of the horizon mirror on to a distant landmark, being careful to choose a distinctive feature. Holding that image steady in the telescope, slide the index bar around the limb until a second and equally precise landmark registers in the horizon mirror—the lower half of the field of vision through the telescope. Allow the vernier screw to engage and make the final adjustment so that the landmarks appear exactly over one another.

As the boat may be rocking, it is important that the observer must twist his body to counteract the boat's movement and maintain the plate quite horizontal at all times. It is as well to take two or three sights each time and take the mean of the three readings to check for accuracy. Once again three bearings should be taken of three landmarks, which should be between 30 and 60 degrees of each other.

To transfer these bearings on to a chart, take a transparent protractor and mark the bearings on the matt surface by drawing from the requisite readings on the edge to the centre hole. Place the protractor on the chart and move it around so that the three lines, or their extensions, pass through the appropriate landmarks, mark a dot through the centre hole and that is your position. If your chart scale is too small, it may be necessary to transfer the bearing lines from the protractor to tracing paper, then extend the lines sufficiently to cover the landmarks, adjusting the tracing paper so that the correct position is established.

DISTANCE AND BEARING BY SEXTANT (p 146)

If the height above the sea level of a landmark is known, by using the sextant in a vertical plane it is possible to obtain the angle between the highest point and the water level. This can be equated to a distance and will give what is known as a 'circular position line', the radius of which, using the observed height as centre, is arrived at by the following equation. Obviously, the radius is also the distance from the object, or more correctly, from the centre line of the object projected to the water level.

$$\text{Distance in miles} = \frac{\text{height of object (ft)} \times 0.565}{\text{vertical angle in minutes}}$$

The answer being easily converted to metres if required.

Immediately after taking the height reading, take the bearing of the landmark and this will give you a bearing and distance fix. If the height of the two landmarks is known, taking the vertical angles of each will give two circular position lines, their crossing point being the position of the observer.

We have several times mentioned thirty degrees as being a suitable angle between two landmarks. This is not a fixed requirement, but it will eliminate the possibility of producing a bad triangle of errors or 'cocked hat'. This sort of angle will ensure a clean-cut crossing point and this is necessary. Taking an extreme example, if we were to sight on two landmarks which were exactly 180 degrees apart, we would end up with a straight line joining them, our position being somewhere on that straight line, but where? Even if we took a third bearing on another landmark we would have only one effective cutting point and a slight inaccuracy in that third reading could move our position considerable distances one way or another along that original line.

Remember, one degree is equal to a distance of one unit in every sixty units of distance from the point of origin. Therefore, if our position were approximately 600m offshore and we were only two degrees out in our readings, our fix would be 20m out of position, one way or the other, a margin of 40m! By ensuring that the position lines are between 30 and 60 degrees apart, even if they do not cut at exactly the same point, the resultant triangle of errors or 'cocked hat' formed by the intersection of the lines will confine the position of the site. The size of the cocked hat will be determined by the accuracy of the readings.

Another point to be borne in mind. Frequently inexperienced surveyors will try to take a fix on the near-vertical line of a cliff or rocky promontory that juts out to sea. This will obviously give a distinctive position to sight upon through a telescope, but from a point some little distance away from the original one, the outermost projection of the cliff could be a very different point to that previously sighted upon. Try it for yourself sometime; just walk along a beach and note what appears to be the most outward part of the promontory, walk along further and you will notice that you can see further round the cliff as you have gone farther from it, but the edge will still appear quite distinctive. Then look at a chart and try to decide just whereabouts on that promontory you should take your bearings! Similarly, but not quite so confusingly, if you sight on to a round tower, unless you line up with the central line, your position line will have a different register point as you move round and see the tower from different positions.

SURVEYING (p 150)

So far we have considered the problem of locating and relocating a site from the seaward aspect, but the time will come when it is desired to plot the site on a final chart. Here it becomes advisable to use methods more accurate than those obtained with a sextant from a bobbing boat. Particularly on harbour projects, many bearings will be required covering a number of positions and these are best taken from the shore. The ideal instrument is the professional theodolite, but if one of these is not available, a home-made instrument as described in Chapter 5 can be used and even from that readings down to a fraction of a degree can be obtained by estimation, or the sextant, with its accuracy in minutes of a degree, might be used.

But a professional theodolite can give an accuracy down to one second of arc by direct reading with no estimation at all. Such accuracy is only usable in mathematical surveying and certainly is unnecessary for the present purpose, for apart from anything else, it is practically impossible to draw a line within an accuracy of one minute let alone one second and even if it were not, we are unlikely to find many protractors that can register, without estimation, a reading better than half a degree.

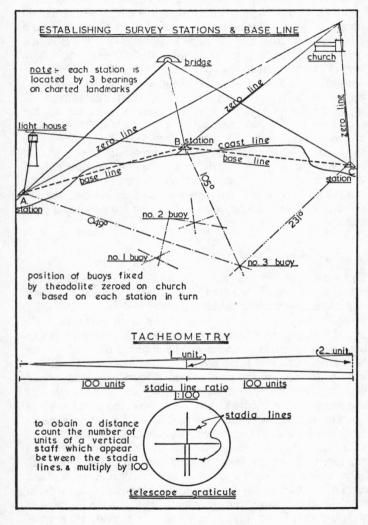

However, the mere fact that such accuracy is possible will probably make the observer take just that little more care in

Locating and pinpointing the site

measuring his angles and that will all be to the good.

Whatever your choice or availability of instrument is, the actual survey is begun in the same way. The first necessity being to establish a base line.

BASE LINE

This is an imaginary line joining a number of points or 'stations' from which are taken all sights of the area to be surveyed. The size and topography of the site and adjoining coastline will dictate the number of stations to be established, but there will never be less than three. Ideally these should be in a straight line and evenly spaced, but neither is absolutely essential and in many cases may not be possible. But it is advisable that they should be within sight of each other. The author's group has found on doing harbour surveys that it has generally been possible to establish our base line along the beach fronting on to the site and from there we have had a clear view of the whole field of operation.

SURVEY STATIONS

The positioning of each station is critical. It must be accurately located and marked in such a way that you may return to it innumerable times with the absolute certainty that your instrument is re-established in precisely the same spot. Each station must have a clear view over the whole area of the site and it must be possible to see at least one adjacent station. Even more important it must have a clear view of good landmarks of its own that also appear on naval or survey charts. This latter point is necessary if the survey is meant to be an extension of a land survey which is based on existing land charts or maps. If the survey is only aimed at plotting the positions of various underwater structures relative to each other, then any base line will do, without reference to existing land features. But there is little point in carrying out a survey if it is not to be tied into the local surroundings.

Choose the positions of your stations with an eye to levelness, accessibility, evenness of height above sea level, and with a free field of vision. Do not, however, do as we did at Nora on our first visit. We established one station in such a position that we had to take some of our bearings overlooking a military base! Within a couple of minutes of setting up the theodolite, we had half the Italian army bearing down on us with bayonets fixed!

Having selected the position of the first station (we will call it 'A'), set up the tripod as firmly as you can (it is assumed in what follows, that a professional theodolite is being used), treading the legs well down into the ground and using the spirit bubble level that you will find in the top of the tripod to ensure that it is quite level. When you are quite satisfied that this has been done well, drop the plumb bob from the underside of the tripod head and mark the precise point on the ground over which the tripod has been centred.

This marking is the key to successful relocation of the tripod on future surveys so great care should be taken. In soft ground, drive a long spike well down so that its head is flush with the soil, paint this white and put a small nail into it to mark the exact centre point. A few stones piled in a cairn will help future location. If positioned over rock, drill a hole into it with a star drill and plug with a 'Rawlplug' or wooden dowel into which a screw or nail can be driven. Sometimes a masonry nail can be driven straight into rock. In each case put a dab of white paint to mark it more clearly. Check your tripod again for level and mount the instrument on it in accordance with the maker's instructions. Having done this remove the plumb bob and check the positioning of the theodolite by using the built-in 'optical plumbline' to ensure that it is correctly sited.

On looking though the circle eyepiece you will be able to see the 'horizontal circle' reading. By turning a small milled knob, you will be able to revolve the circle until a zero appears both in the main scale and micrometer scale 'windows'. Lock it there and rotate the head of the instrument until your main landmark appears in the telescope. Unlock the circle clamp at that point and you will have 'zeroed' your instrument on that landmark or heading. Now rotate the instrument until the next landmark comes into view and check the horizontal circle reading, using both scales. This figure will then be the angle between the two landmarks measured in degrees, minutes, and seconds, as measured at station A.

Locating and pinpointing the site

Rotate the head still further in the same direction until another landmark appears in the telescope, take the readings of this and for a double check repeat with a further landmark if possible. Now lift the whole paraphernalia and move along the base line until you reach the next position chosen for a station (B). Repeat the operation as before and then go through the process once again at station C. So far, of course, no actual surveying has been commenced, but the base line is now established preparatory to the actual site survey. These stations can now be drawn in the chart because all will have at least three sets of bearings which are related to landmarks shown on the master map or chart. The various stations can be fixed using a protractor as described above. The drawing on p 150 will illustrate a typical layout.

An additional check may be made with the theodolite as a direct reading, which is not possible with the other instruments we have discussed. This is to locate a point as a bearing and distance from another, eg station B from station A. This is known as tacheometry.

TACHEOMETRY (p 150)

Most modern theodolites provide facilites for tacheometry and give accurate direct readings to within a quarter of one per cent, or perhaps it would be wiser to say the instrument is accurate to that degree—the final outcome does depend upon the observer. To enable distances to be estimated, an additional pair of short horizontal lines are marked on the object lens of the telescope. If you look through such an instrument, you will see the normal 'cross hairs' which are used for sighting, and in addition you will see the 'stadia lines'. These are the two horizontal lines above and below the main lines. These have been very carefully calculated to subtend a triangle, the apex of which is at the centre of the instrument. This is an isosceles triangle, the base of which measures one hundredth part of the height. Obviously, the visual lines joining the base at the graticule to the centre of the instrument, can be extended visually to any distance and the ratio of the base to the height will still be 1:100. As the 'base' of the triangle is in the vertical plane, all that is necessary is to sight on to a vertical feature of a known height, multiply that height by 100 and we have our distance from the vertical object. In practice, we use a 'vertical staff' which is a telescopic wooden frame on which coded numbers are very clearly marked. These represent feet or metres and can be clearly seen from a distance.

Therefore by sighting on to a ranging pole or vertical staff, we can count the number of divisions appearing between the two stadia lines, and multiplying this by 100, we have our distance. When using a ranging pole (and so saving having to carry a vertical staff for possibly only this one purpose), I find that it pays to sub-divide the uppermost white sub-section of the pole into tenths, lining these clearly with black-painted lines. This reduces the amount of estimation needed, for inevitably the stadia lines will never or rarely coincide precisely with two whole sections of the pole and it is then necessary to estimate what proportion of the segment has to be multiplied by 100 to arrive at the full distance. The observer merely has to count the number of full (feet or 50cm) coloured sections plus the number of tenths of the marked one and multiply, for instance, 1.7 or 2.4, by 100. When sighting on to the ranging pole ensure that a full coloured section starts at the lower stadia line and the upper stadia line will then cut across one of the sub-divisions.

Should the distance be such that the ranging pole will not cover the space between the two stadia lines, count the sections appearing between one stadia line and the horizontal line of the graticule, double this and multiply by 100. In an emergency it is possible to sight on to a vertical feature, for instance a door post, which just fills the gap between the stadia lines, and at a later opportunity measure that feature, and so work out the actual distance. We have even done it on a team member— soles of feet to left nostril!

The above notes are only true if a reading is taken with a horizontal line of sight. If inclined readings are taken, the equivalent horizontal and vertical distances can be worked out as follows:

$$\text{Horizontal distance} = 100S \cos^2 V$$
$$\text{Vertical distance} = 100S \tfrac{1}{2}\cos^2 V$$
$$\text{Where } S = \text{Stadia Reading}$$
$$V = \text{Angle of inclination.}$$

Locating and pinpointing the site

The angle of inclination is obtained by taking the vertical reading on the theodolite of the object from the station. By siting stations as near the water level as wisdom, tide, and prevailing winds allow, the need for making this correction on the actual site survey can be eliminated and for all practical purposes, within the accuracy to which we are likely to be working, it is as well to treat all readings from a station within, say, 2m of sea level, as being horizontal. In any event, except on very shallow sites, ranging poles are not likely to be in use and the tacheometry side of the theodolite is not going to present problems.

The accurate setting up of base line stations is a most important function and a two-man team, if possible under the leadership of someone with experience of survey work, should be assigned to the operation. If by mis-chance, no really competent man is available for the job, double up on the man who is taking the readings, ie have two men on the theodolite and one on the ranging pole. In this way, each zeroing and reading may be double checked. Nothing is more annoying to all concerned than to find, on return to base, that one reading is obviously in error. There is no alternative to having the team return to that station and recheck their readings, if necessary re-laying the marker buoy that is the subject of the check. All very time-wasting and infuriating to all. But there should be no excuse for lack of expertise in the very simple aspects of using a theodolite that we require, for they can be learnt and practised at home, long before the expedition sets out.

While the theodolite team has been setting up its base line stations, others in the group may have been savouring the pleasures of diving and locating finds—much to the chagrin of the survey team—so it makes for harmony if a rota can be established with all members taking turns at the less interesting and usually either blazing hot or freezingly cold stint on the theodolite. The underwater team will by now be buoying the area and each buoy will relate to a numbered find or structural detail. It should be possible for the diving team leader to prepare a rough sketch map of the layout of the buoys, indicating on it the appropriate numbers, for use by the survey team, for obviously they must relate their sightings to known objects/buoys.

The method of doing the actual survey is very simple but does require conscious effort to ensure that it is done properly. It is far too easy to make a silly mistake, which at best, will be wasteful in time. Having set up the instrument at station A, it is necessary to 'zero' the horizontal circle, in other words, set the horizontal circle to a zero heading on a known feature or magnetic north. The former is best if there is one outstanding landmark that can be clearly seen from all stations, for this will undoubtedly be quicker to set up each time. However, theodolites are provided with magnetic compasses which can be coupled to them and with these magnetic north may be located. But they are what is known as 'tube' compasses and consist of a tube in which a compass needle is located. This has a 'locking' device to stop it swinging or moving during transportation and needs to be freed when actually required. Since it is in a tube, it has very little room to swing and the compass usually comes into operation when within one degree of magnetic north. Outside this radius, the needle firmly jams against one side or the other of the tube. It is therefore necessary to be able to swing the instrument to within one degree of magnetic north, using some outside agency as a guide. This takes time and experience, because even when within the one-degree limit, very delicate adjustment on the vernier gauges is necessary to line up correctly.

Depending on the method chosen to zero the instrument, the first step is to line up the graticule accurately on to the landmark or the compass needle on to north. Then release the lock on the horizontal circle and with the adjusting knob (usually covered with a hinged lid to prevent accidental movement) revolve the horizontal circle until both main and secondary readings are at zero. Having done this, cover the adjusting knob and lock the clamp on the circle. This will keep it permanently (until reset) zero'd on that particular feature.

Next traverse the instrument over the site area, noting the reading on the horizontal circle at each buoy. These will appear as direct readings in degrees and, in a secondary 'window', both minutes and seconds. The figures noted will represent the degrees of the angle formed by a line from the instrument to the original zero point and a line from the instrument to each buoy. It is best if you can zero on to such a point that all

Locating and pinpointing the site

subsequent readings are taken in a clockwise rotation as then bearings will get progressively larger each time and this makes for easier readings.

An important point to note here: do be very careful to check each reading and associate it with the correct buoy. From station A the buoys may appear in a numerical sequence of 1, 2, 3, 4, etc, but from station B, farther along the beach, they may well appear as 3, 1, 2, 4! This is especially true if the area being surveyed is a narrow rectangular one with its narrow end facing the shoreline—so do take great care here.

When all buoys have been surveyed, the bearings are transferred to a chart by placing your protractor over each station in turn for each buoy. Zero it on the feature originally used to zero the theodolite and draw in the bearings with faint pencil lines. In this way the three lines for each buoy will form a 'cocked hat' and the location of the buoy will be central within it. It helps to erase the faint pencil lines used to locate each buoy before plotting the next; this saves possible error. When all buoys have been marked and numbered, it is only necessary to joint them together to get the outline of the subject. In the case of a harbour or large underwater structure, it is frequently only necessary to position markers at the salient features of the structure, ie the ends of walls, angles, apex, etc, with possible intermediate markers on long straight runs of walls or roads.

So far we have taken the simplest of exercises in plotting a site and it can be said that this fully covers the requirements in non-tidal waters such as the Mediterranean where most harbour works are likely to be found. But where tides are large, the problems discussed in Chapter 4 present themselves and must be overcome.

9
RECORDING UNDERWATER FINDS

For the first time we have now reached the stage when all your underwater skills will be necessary, where all the practice time spent in swimming baths and open water in learning underwater techniques begins to pay dividends. The relatively simple tasks of finding the site, locating it, and even surveying the major features have all been done. Now it is necessary to get down to detail work, completely underwater and in visibility that may not be of the best. Divers eventually become cold and tired, whereupon efficiency falls off and results suffer. But this must not happen if the job is to be completed properly. To minimise this we must improvise equipment and devise ways of doing things in the shortest amount of time with the minimum effort, but still with accuracy.

The value of a site is exactly what the divers can produce from it. On land, a Bronze Age burial mound ploughed by a farmer may produce a few, probably broken, pots or implements of some monetary value but little else. Were the same mound to have been excavated by archaeologists, what a different story there would be. That is why, in land archaeology, the rule is to hasten slowly; every artifact is carefully noted, measured, related to its surroundings, photographed, and recorded before it is even moved. For within its context, a small find can tell a great deal; take it out of context, and it tells us nothing.

At the time of writing, speculation is going on as to how best to handle the *Amsterdam*, mentioned earlier. We know

Recording underwater finds

that by putting a mechanical grab down into her we shall be able to pull up a lot of articles of interest and even possibly of value, but they will be out of context. Do we put a cofferdam around her strong enough to keep the seas out for perhaps a year or two while the contents are carefully dug out, as in a dry excavation on land? This would certainly be a vast improvement on the former idea, but might it not be even better to make a cofferdam of such strength that the whole vessel, hull, cargo, and all, can be excavated and so present us with a complete picture of her construction, how the cargo was carried, what the cargo consisted of, where and how her crew lived, probably even to the extent of finding out what they wore and ate. Time will tell and no doubt funds—or lack of them—will play a deciding part in this, but obviously the more detailed the excavation, the better.

Any underwater site must be treated in exactly the same way. Every find must be accurately recorded by measurement, drawing, and photography before being moved. I have stressed before the need to record before moving anything. I would go so far as to say that nothing should be moved unless there is someone supervising the operation who has archaeological training. It is therefore essential that an archaeologist should be on the team of every expedition. Even if it is not possible for him to dive, his supervision from above can be of the utmost value. However, having said that I do find myself with certain mental reservations. I would qualify that statement to some degree, although others could possibly violently disagree with me.

An archaeological site on land has built its progressive 'layers' over long periods of time; one development is superimposed upon another. Even the most violent efforts of men can only reduce to rubble what was once built at that place, but the remains will still be in the same place and over a period will merely form another layer. Each of these layers represents an era which can be accurately dated by the most prosaic of finds that may be left within. Pottery, glass, metalwork, and coins are just a few of the more obvious clues. Their presence indicates not only the time span, but also the type of people that dwelt there, what their work was, their religious beliefs, and general ways of life. Each little find fits into a pattern of the site and that site into a pattern of archaeology which together adds so much to our knowledge of bygone times.

But the key here is that the artifacts have arrived in respective positions due to the act of man. I do not mean everything was carefully placed in a special place, but that man's desire to change, to improve, to destroy, or just to accumulate debris upon which he was happy to live and even rebuild, all came about over long periods of slow change by or through man. Therefore everything accumulated in positions relative to the era in which it was created.

But take our wrecks. In Chapter 7 we discussed some of the ways in which wrecks occurred. Almost without exception, they had a violent end and even when on the bottom, they were often still at the mercy of the seas. Their contents were strewn across the ocean bed from the point of impact, on rock or by bomb, and were scattered still more by the disintegration of the ship once it had reached the bottom. In the swell of shallow waters, there are many instances where the contents have been washed from inside the vessel right out of and under the hull! Their significance is hardly relative to the position in which they are found.

Although it has been known for one or more wrecks to have been found in the same spot and even on top of one another, this is an unusual set of circumstances and is still very different to the layering of a land site. The finding of one or two dateable objects can be sufficient to date the wreck and even to identify her. The wealth of other objects tells us even more about the ship and those who sailed in her, but it does not really make much difference if a belt buckle from a sailor's coat is found at spot A, or spot B 3m away. It does not matter too much if a cannon is found lying pointing east–west and another is found pointing north–south, some 10m away! Provided that they are carefully recorded and their positions marked for further searching in the area before being lifted, that they are properly treated for conservation after they are lifted, and that they are properly reported, so that all interested parties can learn about them—that in my opinion should be sufficient.

It is impossible, or at least unwise, to generalise in matters appertaining to archaeology, as it is indeed with anything else,

Recording underwater finds

so what I have said above must in turn be qualified. The fetish of precise recording of every find that is made on the site of a scattered wreck, as propounded by some, is time-wasting and unrewarding; it is also something I believe to be preached more than to be practised!

But on the other hand, a deep wreck which has been relatively undisturbed by wave action such as the one off Grand Congloue near Marseilles, the Cape Gelidonya wreck of a Bronze age vessel, the Mahfia and Albenga wrecks (see bibliography for details) warrant far more detailed examination. Because, in such cases and even in *Amsterdam*, we have a moderately intact wreck. It is only by a detailed 'layering' type of excavation, such as is carried out on a land dig, that their real value can be fully established and, in such circumstances, I believe it is our responsibility to carry out our work with the care and precision of a land archaeologist.

In the case of harbour works, large or small, we are recording by and large very much bigger objects than are likely to be found in a wreck, ie walls, breakwaters, roadways, etc. The importance of correctly establishing their relative positions cannot be over-emphasised, but here again, I submit, not to within a matter of centimetres, for once again the position of an individual block of stone may well have been dictated by earthquake, storm, and sea swell, not by man. It is more important to try to establish the exact line of the original structure by an intelligent appraisal of the present location of a mass of stone blocks, by endeavouring to estimate how they arrived at the present resting place and what the original structure may have looked like, than it is—in my opinion—to record precise details of their exact position and dimensions. Page 182 perhaps illustrates this point, for here we have a wall from which one stone block has fallen and is resting at an angle to the main line. To me, it seems quite unnecessary to record the angle of its present resting place and its precise location. It would seem more sensible to prove by measurements that the stone had fitted into the gap in the wall and so confirm that it had come from there. Once this is done, it becomes part of the wall and needs no further detailed recording. To record the dimensions of each block would not only be time-consuming but would be wasteful. After all, the masons who built that particular wall followed the then current practices for the building of walls, etc. These can far better be examined on one of the numerous land sites that have been excavated.

On the other hand, I accept that on occasions the main outlines may not be evident in a visual survey; possibly visibility may not allow enough to be seen at one time to form a picture of the whole. In such circumstances the plotting of locations of individual blocks, meaningless at the time, may become meaningful when pieced together on a drawing board like a jig-saw puzzle.

On other occasions, the distinct outlines of a building or room may be clearly seen and it is incumbent upon the observer to carry out a detailed layering and recording excavation as he works his way down through the mud and sand which covers the area.

So the essential methods of detailed survey are of vital importance to the underwater archaeologist and should be used in full whenever the situation demands their use, The newcomer to the work should treat every situation as requiring the full 'treatment', until he has reached the more knowledgeable stage of being able to determine for himself the extent to which they should be applied. Also, if visibility is poor it is better to err on the side of caution and make careful detailed notes rather than suffer for the sins of omission.

So let us now see how to set about the recording of details on a site, utilising the equipment described in Chapters 4 and 5.

TRIANGULATION (p 163)

The most widely used method of positional recording under water is that of triangulation, for with this method, precise locations of objects can be established even in conditions of poor visibility or where parts of the site are separated by large boulders or blocks which prevent direct visual recording.

Triangulation is based on the principle that if all three sides of a triangle are known then the position of each corner is known. So by forming triangles all over our site, all stemming from a common, or known, baseline, the vertex of each triangle will be a precise point which will be related to the baseline. This point can be transferred to a chart by relating the lengths

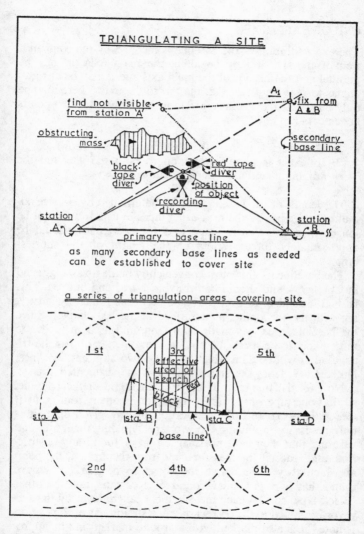

of the sides of the triangle to a scaled-down baseline, as radii of arcs. So the first step is to establish a baseline.

USE OF BASELINE

The exact location of the baseline is dictated by the configuration of the site; it may be along the main axis, it may be parallel and to one side of the main axis and it may be at right-angles (or any angle for that matter) to the axis. There may be only one or, if the site is larger, more than one. But two points are essential:

(1) The exact length of the baseline must be known.
(2) Its exact location in relation to both the site and the adjoining land mass must be known.

The length of the baseline will be determined by, as a maximum, the lengths of the measuring tapes in use. It must not be longer than the shortest tape and for preference should be approximately three-quarters of the length of the shortest tape.

The baseline itself is most conveniently made up as suggested in Chapter 4, and should be firmly anchored and buoyed. This is then located in respect to the adjacent land mass by surveying with the theodolite, as described in the last chapter. Since the baseline rope was made 50m long, and as 30m tapes are most likely to be used, the obvious baseline length for the triangulation will be 25m which will give two adjacent baselines, if necessary, from the one black-and-white segmented rope.

Now fix the free end of a tape to each end of the baseline, so that one tape runs from point A and another from point B (see p 163). To record the location of any given spot, it is only necessary to unwind the two tapes, holding them taut and ensuring that they have not snagged on or round any obstruction, and read off the measurement from the tapes at the point at which they cross above the object to be marked. I have found that it pays to mark one reel 'Red tape' and the other 'Black tape' so that each measurement taken down by the surveyor is applied to the correct baseline station. As each position is noted it is also marked with a ground marker and its appropriate number recorded with the measurements.

PROJECT	Bythia 1969		
DATE 15 8 69	AREA 'Jug'	SURVEYOR G Stewart	TEAM B

Marker Number	Description of find/point	Station readings	
		Red Tape A	Black Tape B
01	neck of amphora	01 45	24 02
02	floor tile, hexagonal	02.51	23 75
03	corner of low wall (SE)	03.05	23.00
04	corner of low wall (SW)	03.55	23.07
	etc		

CONDITIONS: Bottom Sand   Visibility 30m   Current/tide nil

GENERAL OBSERVATIONS:

No problems, many sherds still to be measured. May need dredge later.

USE OF TAPES

If considerable triangulation is anticipated, it pays to paint the record sheet outline on to the drawing board or at least on to a sheet of plastics material so that it appears through successive sheets so facilitating the recording of each object. See page 165 for a typical layout.

The following suggestions will make the task easier. Try to work in a clockwise direction from the baseline; this will ensure thorough coverage of the site and will help the draughtsman when he comes to transfer the readings to his chart. Use a team of three. One acts as surveyor, records all measurements and affixes the ground markers to the points he is recording, and, in fact, is team leader for the exercise. The other two each have a tape and that in itself can be quite a handful! It pays to unwind only as much as is necessary for a particular measurement; any slack in the tape will only tend to snag on obstructions or foul up on the diver himself. Then swimming with the tape, keep it as clear of the bottom as possible and when taking the measurements ensure that the tape is firmly taut, but do not use brute strength. It is difficult to know how much strain you are applying under water and too much could result in the metal end tags being torn away. If the surveyor can do it, it will help both him and the draughtsman, if he sketches the general plan of the area he is working. No great accuracy is needed but it will put the numbered markers into relation with each other and so make the job easier. Be careful to ensure that measurements are taken in the same unit each time—this particularly applies if a combined metric and feet tape is used.

When the whole area has been completed, the team can turn to the other side of the baseline and repeat the operation. They can then move on to the next section of the baseline and so on.

A problem that is quite likely to occur is that a large boulder comes between a particular point to be recorded and, say, base station A. Consequently, a tape measurement direct from A would necessitate going over or around the boulder and an inaccurate measurement would ensue.

To overcome this, find a point from which the area behind the obstructing boulder can be clearly seen and fix this by measurement from Base A and Base B in the normal way. Call this point A_1 and using A_1–B as the new baseline, continue to triangulate the area. It will not be necessary to buoy A_1 for surveying because its location will have been fixed by triangulation from A and B. As many secondary baselines as are

Recording underwater finds

necessary can be established, but the placing of the original baseline should endeavour to take into account any major obstructions and so obviate the need for secondary baselines.

Back at base, the draughtsman interprets the various tape measurements as radii from the respective base stations and with compasses draws the two arcs, the intersection of which defines the position of each object. He then numbers the objects according to the surveyor's report.

Using a measuring chain should, in theory, provide more accurate readings than a tape, but in practice chains are a complete nuisance, especially if the bottom is weedy, as they will continually become snagged.

USE OF PLANE TABLE

The construction of this instrument was discussed in Chapter 5 and we shall now look at the method of use. As with any instrument depending upon visual orientation, it presupposes that visibility is good.

The plane table is positioned at a strategic point from which the whole of the area to be surveyed can be clearly seen. Its precise location should be established and recorded by buoying and theodolite survey from land. The table should be firmly resting on a stable bottom and should be levelled by using a spirit level. Use additional rock weights if necessary to ensure that no accidental movement can take place. On to the table top fix a clean sheet of plastics drawing 'paper' and make sure that it, too, is free from risk of accidental movement. At one of the corners, probably the near left hand, insert a small nail or pin, again ensuring it is quite firm. Place the sighting instrument on the table so that the small vertical 'V' at the end of the drawing edge engages around the nail. This nail now represents the station and becomes the pivot point for the instrument. It follows that any line drawn along the straight edge will radiate from the nail and therefore from the station.

Another diver equipped with a ranging pole and markers, swims into the site area and holds the pole vertically over any objects that he finds, having marked each in turn with consecutive numbers, noting a brief description of the object against the number, on his note board.

The diver at the plane table sights through the instrument on to the vertical line of the ranging pole, then draws a pencil along the straight edge, marking the underlying sheet. Against this line he puts the number of the object. It is as well that the diver with the ranging pole should hold up the number for the surveyor to read, before placing it on to the object; this will ensure that each applies the same number to a find.

When all the objects have been recorded from one station, the plane table is moved to a new station, which is again clearly established and recorded. A second nail is inserted in the table top at a point scaled to the actual distance from the first station. Thus a 'base line' is established, as in the triangulation method. Both nails can be located prior to the start of the survey, if the site has been previously inspected. The whole process is now repeated, the ranging pole being held over the numbered objects as before, the numbers being confirmed visually to the surveyor and another series of lines drawn, each in turn being numbered. The intersection points of like numbered lines represent the positions of the objects.

LEVELLING BY PLANE TABLE (p 169)

At the same time, providing the table is perfectly level, it can be used as a datum height. When lining the vertical graticule of the instrument with the ranging pole, one can also note where the horizontal line cuts across the pole and a plus or minus height differential can be registered. Note this against each numbered object and you have all the information necessary to establish levels throughout the site. If it is intended to check levels, it is wise to locate each of the plane table stations at relatively high points. In that way, most of the readings against datum height will be higher on the ranging pole or levelling staff and will therefore indicate the distance below datum of the point being measured. This may sound confusing, so I will explain further. For practical purposes it is not necessary to know the exact height of the graticule above the sea bottom, although at a later stage it may be of value to know the exact depth at which it is situated.

We know the table is level, so we will consider that the height at the graticule is zero metres. If, on sighting through

Recording underwater finds

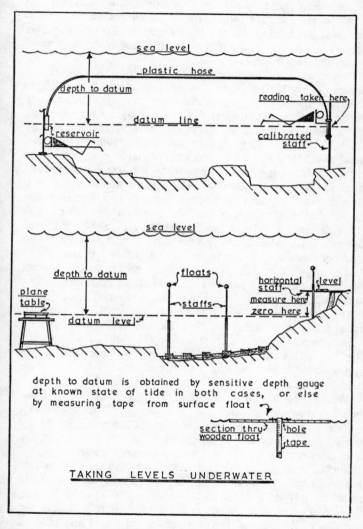

the instrument we see that the horizontal wire of the graticule cuts the levelling staff at 1m, it must be obvious that the foot of the staff is 1m below the graticule or datum level, if it cuts it at 3m, it means the foot of the staff is 3m below datum and so on. If the level of the object was exactly that of datum, we should see the top of the object and the foot of the staff meeting on the level of the cross-wire.

When we come to draw up a cross-section of the site, we arbitrarily establish a datum line and then measure, down from that line, the reading appropriate to a particular object to establish its level against the datum line. By joining up these levels, we have a section of the site showing all the relevant levels. This may help to explain the need to establish the plane table at a height at least on the level of the highest point to be surveyed, for if it were below some of the levels, it would not be possible to see the levelling staff when it was standing upon those particular points. At a pinch, if one or two objectives are, in fact, higher than datum and this is unavoidable, we can still get a reading. Hold the levelling staff vertically between the object and the plane table, and use another staff as a horizontal member (use bubble level) resting on the top of the object and touching the staff. Note the staff reading at the contact point. Check the level against the plane table graticule reading and subtract the lower from the higher reading. The difference is the height of the object above datum level and must be clearly shown so that the draughtsman will take this into account when doing his cross-section. The difficulty of ensuring that both vertical and horizontal members are true, allows considerable scope for error, so this method should only be used in a last resort.

With the plane table, directional bearings of the objects from two base-line stations (or more if necessary) have been taken without actually noting the bearings as degrees, merely by drawing the lines representing those bearings. If accurately and carefully executed this can save draughtsman's time on shore; he merely lays the underwater drawing over his master plan (see that the baseline has been drawn to the same scale as his chart!), lines up the base-line stations and pricks through the underwater drawing on to his master chart to locate the position of the objects. If visibility is good this is an excellent method of survey, although it takes longer to carry out than triangulation.

USE OF THE UNDERWATER 'THEODOLITE'

This instrument is positioned exactly as is the plane table. The observer again sights through it on to ranging poles held over the objects to be recorded, but instead of a line being drawn to indicate the bearing, it is actually read off the protractor as a bearing of x degrees. The instrument must be zeroed like a land theodolite, either on to magnetic north (ensure here that you are not sitting astride a vast iron cannon) or on to another known point or even on to the second station along the baseline. In practice the latter may prove the better. Whichever is used, it should be carefully noted for the guidance of the draughtsman. Having taken the first set of readings from one station and also levels if required, as explained above, move the instrument to the second station and repeat the operation. Each object will then be tied to a station by a set of bearings and the draughtsman merely centres his protractor on each station in turn, zeroing as instructed, and reads off the bearing and draws it in. The intersection of each pair of lines is the site of the object being surveyed. Quite simply done and pretty accurate.

Both the above systems depend upon two operators and good visibility. Another device we have mentioned enables fixes to be obtained when visibility is not good and even with only one diver if necessary. But I would not like to think that there will be a case where one diver is in operation by himself, for the obvious and very good reasons known to all practical divers.

BEARING CIRCLE AND MEASURING LINE

The construction and use of this was explained in Chapter 5, but I cannot think that it is an ideal instrument. I have certainly not used one myself so cannot speak from experience, but it does seem that errors of identification of objects, bearings, and distances are too liable to make this no more than an approximation of position finding. If visibility is too bad for the use of the plane table or sighting tube, then revert to triangulation every time, for then accuracy is assured.

Even above water, with an accurate instrument such as a theodolite and tacheometric readings of distances, errors can occur. Under water this will almost certainly be the case.

We have covered the question of taking levels both earlier in this chapter and in Chapter 5, where we discussed the use of an underwater 'aqua-level'.

USE OF 'AQUA-LEVEL' (p 169)

With this device, once again levels are not established against any natural feature or water surface, but against an arbitrary datum level. As before this should be slightly higher than the highest point to be measured and for convenience to the diver preferably at least 50cm above that high point.

Two divers are required. One is stationed at the air reservoir end and his sole job is to ensure that the air-to-water interface remains on the datum line engraved or marked on the reservoir. A small variation in this level will be considerably exaggerated at the other end where the bore of the tube is very small in comparison, consequently readings will be inaccurate. Even at the reservoir, relatively small amounts of air will be required to maintain a correct level, so treat this with care. A boring job for the diver and very cold to carry out—but necessary nevertheless.

The responsibility for the actual taking of levels rests upon the second diver, particularly in conditions of bad visibility, when he might well be out of sight of his colleague. In such conditions, it might be necessary to double up on each diver for safety's sake. This diver must select his spot heights, mark them with a ground marker, note the description of the object, and then take his level.

To do this he simply stands his measuring rod, levelling staff or whatever he is using, vertically upon the point to be measured. Holding the open end of the tube pointing downwards, he notes the position of the air-to-water interface and, against that point, notes the measurement on the staff. This measurement, whatever it may be, represents the distance below the datum point at which the base of the staff is situated and therefore the top of the object for which a level is sought.

It is better to release the clip or other closure at the end of the tube, at a fairly low level, raising the tube slowly until the correct level is reached. If opened when the end is too high, air will spill out and the chap at the reservoir will have to work overtime!

The suggestion to raise the tube slowly may sound unnecessary, for obviously there can only be one level at which the water and air will meet. But should the level to be recorded be very much lower than the original datum level, the interface may well be above the head of the diver and he will have to rise slowly until the interface is at a more convenient level relative to his face for him to take the reading. The fact that initially he may have taken the tube well below the datum level will mean that air has been forced out of the tube into the reservoir and upon the diver rising to take his reading, more air may have to be fed into the reservoir to maintain the correct levels. From the foregoing, it can be seen that in a particularly uneven site, even with careful control of the 'free' end of the tube, constant readjustment may well be necessary. This mitigates against the idea of blowing air into the tube from your mouth.

This system enables accurate readings to be taken over a wide area without visual reference to the datum point. With practice it is not difficult to carry out a large number of readings within a short space of time. Once again, it will repay the effort if the survey is carried out in a systematic way, working from one end (datum) of the site and gradually traversing at wider radius until the limit of the tube extension is reached. The tube will float well above the diver's head and so will neither become an obstacle in itself nor will it foul obstructions on the bottom. However, it is a good idea to have a spool upon which to wind the tube on completion of work for 30m of plastics tube can live a life of its own if allowed to, especially when one is trying to coil it up preparatory to swimming back to the shore or attendant boat.

On completion of the survey, it is only necessary to take the depth of the datum line from the surface by a plumb-line or even possibly a 30m tape, then by relating all other points as a plus or minus reading against datum, accurate depths of all points can be obtained. If the actual depth of the site is not considered particularly important, the cross-section of the site can be described as mentioned before, by relating all levels to an arbitrary datum line.

INCLINOMETERS

I certainly have had no experience either of making or using one of these instruments, but in conditions of good visibility I have no doubt such a device could be used. In a simple form (see figure, p 194) I would imagine a sighting device mounted on a pivot to allow movement in a vertical plane. Fixed to the side of this could be a protractor, also in a vertical plane, with the curved edge downwards, and its centre at the pivot point. If a weighted line or a wire with a plumb bob on its end were to be suspended from the pivot point as well, it would always hang vertically downwards. Since the protractor would move with the alteration of angle of the sight device, the vertical line would always register the actual degree of inclination on the scale of the protractor. The device would be worth trying, although I feel that in visibility necessary for its use, it would be as well to use the plane table and do two jobs at once.

So far we have been concerned with the use of equipment aimed at registering the large masses in both horizontal and vertical planes. But assuredly much more detailed survey work will be required. So we need equipment to integrate a find more closely into the geography of the site. For this we turn to the grid prepared earlier.

USE OF GRID

It has been explained how grids may be made and established upon the seabed over the site. The size of grids will depend entirely upon the size of the site or specific area on which we are working. Their size and anticipated duration of use will also decide how they are to be made. A relatively temporary affair utilising angle iron or alloy, with cross wires or cords to sub-divide into smaller sections, will do for a job lasting a short time or where the grid will have to be moved to another portion of the site, whereas a much more substantial unit made from scaffolding pipes, etc would be needed on a complicated wreck site, especially if the site were not level from end to end, but on a slope.

In such circumstances, it is preferable to step down the adjacent grid squares to maintain a reasonable level above the site face at all times. Where this is done, it will undoubtedly help to use substantial vertical members of scaffolding, well driven into the seabed, from which the horizontal members will run at their respective levels, using ordinary scaffold clamps to lock them firmly in place. (See figure, p 68.)

But whatever the construction method, the principle is identical. All that is done is to divide the site area up into smaller areas which are identical in size and shape—the easiest shape being the square.

By use of letters and numerals co-ordinates can be established quickly and simply. Just as one uses map co-ordinates on an ordnance survey map, so we can apply our own co-ordinate references to the grid. On gridding a large site it may well prove most satisfactory to use the letters of the alphabet to indicate the larger squares, with numerals to indicate the metre squares and originating from the lower left-hand or south-west corners of the grid. Again depending upon the size of the grid, it may be found that 1m squares sub-divided into decimetre sub-divisions as secondary figures will be satisfactory. Anything smaller than this will be unnecessary in my opinion. It will not be necessary to use cross-wires for these decimetre divisions, providing they are clearly marked upon the main frame, preferably in alternating black and white segments.

A grid reading could then read, for example, B0613. This would indicate that the object was in the main square B and would be 0.6m to the right or east of the south-west corner and 1.3m up or north from the same corner. Always give 'eastings' before 'northings' in the time-honoured way. Ensure that the grid letter reference reads large and clear; it is best painted on a plate welded or bolted to the south-west corner of each main square to ensure its permanency. This will also be the originating point for the numerical reference for each square.

Once the grid has been laid (and do take great care over this, paying attention to orientation, accuracy, and strength) it will be a good idea to prepare outline grids on underwater paper or painted on to the draughtsmen's drawing boards. The divers can then go down and as they work their way through each square, merely mark the objects upon the already prepared grid plan. This will save time underwater. If the sides of the grid have been sub-divided as suggested, it will hardly be necessary even to take a tape or measure down, for close enough registering can be obtained by sighting on to the frame measurements.

As each object is located it should be marked with a flat-bottom marker or a floating buoy, depending upon whether or not it is intended to photograph the site with vertical shots. If it is possible, this should be done, especially when excavating down through successive layers of a wreck mound, as one may well be doing. In this way a series of photographs will supplement the drawings as the work progresses downwards, each being related to a particular grid by the reference letter appearing in each corner of the main square.

When working within a grid, it may be advisable for the diver to take off his flippers for these can easily get tangled in the cord or wire cross dividers of a temporary grid and make it rather more temporary than was intended! Personal weighting should be adjusted so that the diver is not constantly sinking down into the grid and disturbing both it and the sediment on the bottom.

The actual drawing on the diver's board will depend upon his draughting ability, but essentially should be accurate if not artistic. Each article drawn should bear the reference number of the find marker and a brief note as to what it is. If the find is made of wood, use the skewer type of fixing and pin it carefully into the wood, choosing a point where it is not likely to cause much damage. If it is a metal or stone article, into which a pin cannot be inserted, use the slip-knot of the fixing cord and loop it around some protruding part. If flat markers are being used, a pin or skewer pierced through them will usually suffice, or a piece of wire threaded through one corner and wrapped around the object will do when it is not possible to pin into it.

The grid dimensions referred to above and the co-ordinates used are related to a small site grid as would be used in a

Recording underwater finds

wreck excavation. However, exactly the same system could apply in the survey of an enormous harbour. There objects would not have to be located to within a decimetre, initially probably not even to a metre, consequently the grid could have main squares with sides as long as 100m, and sub-squares of 10m—it would not matter. The co-ordinates would apply in exactly the same way. Of course, in such a large operation, it would not be feasible to use a metal grid frame: a grid of tapes, yellow or white, stretched out over the bottom, would form the framework. As these might be buoyant, it would be necessary to fix them firmly to stakes or poles driven well into the seabed. However, I feel that this is an unnecessarily time-consuming method of search and recording and personally prefer to rely upon individual buoying and shore-based survey. The grid system is, in my opinion, essentially a means of obtaining very detailed positioning of a multitude of small objects within a confined area, and should be limited to this purpose.

The actual detailing of records should also conform to a set pattern. Again there is no one system which is best; personal preference comes into it. So devise a system and stick to it. It should ideally be in the form of a diary noting each day's work, conditions including weather, time spent underwater, objectives aimed at, and objectives achieved. Detail names of team members carrying out specific tasks, equipment used, and snags or problems arising from its use.

Every underwater drawing, chart, report, or reference should be numbered sequentially and dated. These details should be entered into the day book. On surfacing, each diver should be 'de-briefed' while details are fresh in his mind and a full report entered there and then. His drawings and measurements should be gone over so that any discrepancy can be noted and arrangements made for a re-check on the next dive.

A 'Finds Book' is essential. This should contain entries of every find made with its reference by number and location by grid co-ordinate. The name of the diver who made the find should also be noted together with all facts relating to the find, its condition, appearance, and any special interpretation that can be put upon it at that stage.

From the very beginning, a chart should be built up, starting with the adjacent land mass and showing survey stations, etc. As fixes are obtained for various underwater baselines they should be drawn in and as more detailed finds are located and fixed, they too should be entered on the drawing. Even if, at this stage, the work is done without great regard to final accuracy, it will prove invaluable for detecting any obvious discrepancies in measurements or positioning. If the team includes a competent draughtsman he should be given responsibility for the charting and not be asked to do anything else in the way of land duties. In this way he should be able to complete accurate charts as the work progresses and this is invaluable.

Drawings, to scale if possible, should be made of all finds as brought ashore, even if they are apparently small and unimportant. If scale drawings are not possible, sketch drawings with measurements marked against the main features will help. All drawings should be numbered, dated, and filed in numerical sequence. The use of graph paper can help here immensely. A draughtsman will require a good drawing board, tee- and set-squares, protractor, and a set of drawing instruments. It is essential that he has a table for his work which will not have to double as a workbench for de-coking an outboard engine, etc! It is well worth taking cardboard tubes so that completed drawings can be rolled and stored safely. It is also a good idea to fasten a sheet of plastics material to the top of the drawing board so that it can be pulled over the drawings when the board is not in use, otherwise they will suffer from dust and dirt. For the actual drawings I would again recommend the use of one of the plastics drawing materials. Though perhaps more expensive than tracing paper, they will withstand the inevitable blowing around and rough usage they will certainly be given. If a draughtsman is in any doubt as to how to present his drawings or photographs, he can learn this in advance by studying the work of land archaeologists and noting their interpretations. Some of the main points are illustrated in pp 264–266.

As photographs play an important part in any expedition reporting, the subject will be dealt with in more detail in the next chapter, but it might be remarked here that if it is possible for the black-and-white negatives to be developed on site, it should be done. In this way, the team leader will be able to decide immediately whether they are satisfactory or whether

Recording underwater finds

another shot will be necessary. All photographs should be numbered and should bear the reference number of the find and grid co-ordinates (even if these appear in the picture), indicating what the main subject of the photograph is. They should then be filed in sequential order, preferably in a book of their own. This could also contain information relating to exposure, speed, etc of the various pictures taken.

10
THE PLACE OF PHOTOGRAPHY IN SURVEYING

PHOTOGRAPHY CAN be an invaluable aid to underwater work, for it enables detailed examination of an object to be carried out above water, when time is not at a premium and the observer's mind is free to concentrate on interpretation. It also enables non-divers to see how a find looks in situ and so decide on the best method of excavation. But even before this stage is reached, photography can assist in actually finding the objective.

AERIAL PHOTOGRAPHS

The use of aerial photographs over land has enabled many an archaeological find to be made and it offers similar benefits over the sea. Miss Honor Frost was probably one of the first to use the camera from a low-flying aircraft as a means of recording submerged remains of buildings and other structures; from her work a system of photogrammetry has developed which will be discussed in more detail later on. Vertical aerial photographs have helped the writer in work done on the sunken Phoenician harbours in Sardinia, but prospective users should be warned that although a single vertical photograph may indicate the outline of underwater structures, it will tell nothing of the section through such objects. At Nora, an aerial photograph showed a line of masonry running across the bed of the sea and not unreasonably, this was identified as being a mole protecting the harbour. In fact, upon diving on to it, we

Page 181 (above) A numbered and lettered grid forming the base of a camera support frame is resting on scaffold pipe type of grid erected over the fourth-century wreck at Yassi Ada; *(below)* a high pressure water jet, with nozzle made from ABS plastics. A further extension screwed on to the threaded portion adapts the jet for connection to the underwater dredge

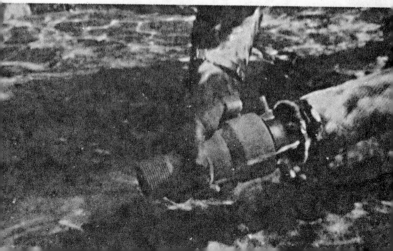

Page 182 (above) An area of the foundation blocks, Bithia. Erosion and weed covering often make it hard to distinguish man-made construction from natural rock. The sixth century BC pottery jug was found in the narrow channel at the top of the picture; *(below)* measurements of this stone block, once part of the wall on the left of the picture showed that it had fallen from the gap and the underside showed the same curved profile as the top course of stone blocks

The place of photography in surveying

learned that it was a paved area of quay forming the edge of the deep harbour; the area between it and the land mass was covered with the remains of buildings and other structures. This did not show up in the one aerial photograph; nevertheless it was a guide to the presence of man-made structures and so was valuable.

BALLOON PHOTOGRAPHY

Aerial photographs need not come from aircraft—the Cambridge University Underwater Exploration Group in 1969 surveyed the Greek site of Elaphonisos using a lighter-than-air balloon. This was just large enough to carry a camera suspended beneath it, complete with a radio control actuator, as used in model aircraft, to operate the shutter release and wind-on mechanism. The balloon was then flown over the site and located by three lines to known anchorage points: from this the precise height and horizontal location could be calculated. A series of photographs was taken with some degree of success although I believe that eventually 'what goes up must come down' applied and both balloon and camera ended in the sea! Nevertheless the feasibility of the idea was proved and, given calm air conditions, there is no reason why great success could not be obtained from this device. Under water, the camera can play its part in recording single finds and in assisting with the actual mapping of a site, its main limitation being the restriction imposed by light and visibility. But by using black-and-white film of adequate speed and sensitivity, pictures may be obtained under quite remarkably adverse conditions. It is not intended here to delve into the very wide subject of photography as an art or science. This is very well covered in books by authors far more competent than the present writer; questions of speed, exposure, light, and colour loss, etc are adequately dealt with in those works. Here we are only concerned with the use of the camera as an aid to surveying.

GRID FRAME PHOTOGRAPHY

We have already mentioned the use of photography with the grid system and the need to ensure that some of the markers used can be 'seen' and registered by the camera in vertical photographs. Many constructions have been devised to ensure that the camera can be located centrally and vertically over a grid square. These can certainly be used effectively and generally consist of a metal box framework which has a square section at its base. This square is similar to that of the grid square and the four corner legs are designed to sit on the four corners of the square. The framework extends to such a height that the whole area of the square will be covered in each exposure. At the top is a device for holding the camera in a lens-down position, which ensures that the camera is level and pointing vertically downwards. A photograph so taken will encompass the square under review and will utilise the sub-squares of the grid to locate any visible find (p 185).

A development of this would be to build into the base of the camera support a fixed grid, so that the whole contraption can be moved from place to place and will carry with it its own grid reference for scaling of finds. But obviously, some system is required to 'lock' the portable grid into the plan of the whole site, so satisfactory location of the area of each photograph upon the master plan will still probably depend upon the original grid system.

The author has not used either of these systems, which were originally developed by George Bass and his colleagues when out in Turkey, but there can be no reason to doubt their effectiveness. It may, however, be asked whether the time taken to move the device each time a photograph is taken, to relocate, level, and plumb, is not so time consuming as to detract from its usefulness.

CAMERA AND PLUMB BOB

However, if this system is not used some other method is needed, for it is obvious that vertical photographs, if they are not to be distorted, must be vertical. A simple idea, but one which needs quite a lot of practice, is to use a plumb-line and weight. Decide upon the height that will enable you to get a photograph of the whole area to be surveyed, which ideally would be such that you photograph an area larger than that which is important, for there is usually distortion around the

The place of photography in surveying

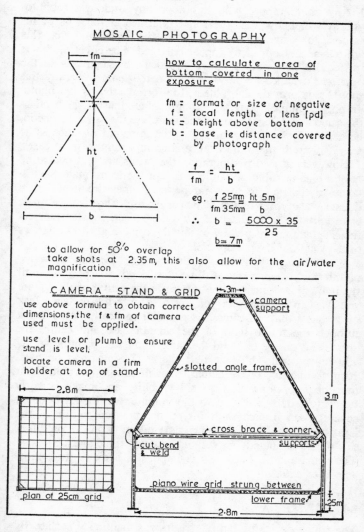

edges of the picture. The depth and underwater visibility naturally has a bearing on this as well.

Any photographer who has done any underwater photography will know that the water interface at the lens of his camera (or camera case) will cause distortion and that this will cause any object to appear enlarged. The effect of this is to require the photographer to focus his camera at the *apparent* distance of the object from the lens. Generally this should not provide a problem, for fortunately, as the photographer is himself looking through a face mask, which offers the same air/water interface, and therefore the same distortion, he need only focus the camera to the apparent distance as he sees it himself; the true distance is unimportant in most cases.

However, in grid photographs, the prints are required to be as sharp as possible, so while it is possible to estimate the distance to the bottom, it is far better to measure it and then deduct one-quarter of the distance. As each shot will be made at exactly the same distance from the object, constant correct focusing can be ensured by setting the camera at three-quarters of the length of the plumb-line.

Fix the end of the line to the camera case and allow the plumb bob to swing free. Adjust your depth and position until the bob is just touching the bottom in the centre of the square you wish to photograph; you are then only left with the task of ensuring that the camera is pointing vertically down. Look through the frame sight of the camera case and centre the cross-wires on the bob and the camera must be pointing vertically downward. The fact that the plumb-line may be fixed 10cm to one side of the actual lens will be unimportant. Paint the bob a bright yellow to aid visibility and make it fairly heavy to reduce swinging to the minimum and a fairly quick coverage can be achieved. If you are not too certain of the light conditions, I would suggest trying an exposure on either side of the estimated 'correct' one, making three in each frame. The cameraman's buoyancy is important here; neutral buoyancy is essential.

3-D FRAME

The frame device of Ian Morrison was mentioned earlier. By means of two photographs of the frame, one for plan and one for elevation, a number of objects encompassed by the frame can be located. Again this is something which the writer has no personal experience in using, but it would seem to be an excellent device for photographic coverage of an area crammed with small objects that require recording, but which numbers and time make impossible to record by the usual 'on site' method (p 70).

MOSAICS

A development of single grid frame photography is the scheme of taking a series of shots to form a picture coverage of an area. The base-line rope suggested earlier, ie with alternate black and white segments, can be used here. Lay this along the line of the area to be photographed and firmly fasten each end. By swimming along this line with a plumb bob and stopping at fixed intervals with the bob resting on the rope, the operator can take a series of photographs which can be orientated and spaced according to the numbered tags on the base line. For best effects, each shot should be taken so that it overlaps the previous one by 40 per cent.

It is then possible carefully to cut out each print, preferably not as a straight-line cut, but following the contour of some object, and join them to make a mosaic of the area. Ensure that the prints are on thin paper to make the joints less obvious. For a really first-class job, having cut the prints to the desired shape, carefully soak them and peel off the backing paper so that only the emulsion layer is left and carefully stick these to a mounting board with a cellulose additive—wallpaper paste is suitable. When they have dried, the joints will be virtually undetectable and a perfect mosaic will present itself. Special stripping papers, which make this somewhat difficult task easier, can be obtained.

The presence of the black-and-white line running through each picture will automatically provide a scale for any object shown. You may find that features at the edges of the prints may not match up exactly, especially if the shots have been taken from fairly high above. This is because of lens distortion as objects are farther from the centre of the picture. The large overlap will help to overcome this to some extent. Also, of course, two adjoining and therefore overlapping prints, will have been taken from camera stations some distance apart. Therefore objects near the adjoining edges will have been photographed from different angles and may not match up exactly. Again a reason for the apparently large overlap, which will minimise the effect. Despite these slight disparities, the overall presentation will be good.

Now an overlap of 40 per cent was mentioned; just what does this represent in distance over the bottom? This can be worked out quite easily if the focal length (pd) of the lens, the format of the negative and the distance above the bottom at which the lens is stationed are known.

The formula is:

$$\frac{\text{pd of camera}}{\text{format of plate}} = \frac{\text{height of lens above object}}{x}$$

Assuming a 35mm camera with a lens of 25mm focal length (pd) with the lens 5m above the base line, we have:

$$\frac{25}{35} = \frac{5000}{x}$$

$$x = \frac{5000 \times 34}{25} = 7000\text{mm or 7m}$$

This figure is based on the camera being held parallel to the base line, for with most 35mm cameras the prints are wider than their height. Were you to hold the camera so that you faced along the line, the format would be less and therefore the area covered would be less, consequently the photographs would have to be taken closer together.

We have also taken the height of the lens above object to be the actual height and not the apparent height, so we must make allowance for magnification and reduce the 7m base by one-quarter. The effective base per shot is therefore 5.25m. If a 40 per cent overlap is desired, the vertical lines of each camera position should be not more than 3.15m apart. The diagram on p 185 may help to explain this more fully.

STEREO PICTURES

If we were fortunate enough to be able to obtain the use of a stereo viewer and possibly even a stereo photogrammetry machine, we would take our photographs as suggested above, but would ensure that a greater overlap is used. This should be a minimum of 50 per cent and, if possible, 60 per cent is better. The effective base line, and therefore centres between stations, can be worked out as above.

Having mentioned photogrammetry, we had better look at the science, for science it certainly is. I cannot do better than to recommend those who wish to really become knowedgeable in this work to read *Simple Photogrammetry* by C. C. Williams. This book investigates all aspects of the work, including underwater, in a very full and detailed manner. Mr Williams has had vast experience in land photogrammetry and has worked with Miss Honor Frost on her survey of sites in Syria. Their combined results are of inestimable value to the serious aspirant to photogrammetry. It is not light reading, by any means, but well worth persevering with.

PHOTOGRAMMETRY

The best way to carry out this work is undoubtedly to take good stereo photographs, get hold of a stereo photogrammetry projector and then sit back in comfort with your eyes to the viewer. A little dot of light traverses the one (combined) picture you see through the viewer and all you have to do is to trace the dot around the outline of whatever you wish to record. Every now and then the dot will vanish, apparently going under some object in the picture—quite fascinating! Upon judicious twiddling of various knobs, the dot will reappear and will traverse over the object and, with more twiddling, will move down the opposite side. In effect, the dot of light can be made to move across the photograph as though it were a real dot of light traversing a three-dimensional model. All the time you are being fascinated by the dot, the machine is doing your drawing for you on a piece of paper to one side. To me this is the ultimate in surveying—but well beyond the means at my disposal and probably yours, so we have to do it the hard way. Twenty-odd years ago, I worked as a set designer in a film company and one of the jobs that used to come up was to project plans and elevations into a drawing so that the art director could see what the set would look like through the camera and when projected on to the screen. We set up the camera position on the plan, drew in the camera angle so that we could find out just what parts of the plan would appear from that camera position, repeated the operation on the elevations of the set and then arbitrarily established the plane of our picture at right-angles to the centre line from camera (eye) to vanishing point.

We then drew in the lines of sight of various parts of the set decor, ie from the object back to the lens, and the points where they cut the plane of our picture would be the position, height, etc, in which they would appear on the screen. This was not hit-or-miss guesswork but precise positioning and the end product would reproduce the set exactly as it would appear on the screen. This was far cheaper than building a model of the set just to get the effect of one angle of shot.

Photogrammetry is a similar exercise but in reverse; we start with the picture and develop the plans and elevations from it. It is not necessary to have stereo pictures for this type of photogrammetry, or even more than one picture. Providing the picture is clear, preferably has a scale or object of known dimensions in it (this is not essential, but it does remove an area of possible error), but certainly has an object or marking that defines a square or rectangle—we can start.

If one were to photograph a grid from a vertical position centred over it, the plan that would appear on the print would show a pattern of squares, but those nearer the edge of the print might be distorted, with the outer lines forming gentle curves convex to the centre of the print. For that reason, objects at the extreme edge of the picture should be ignored; concentrate on those nearer the centre which are correctly reproduced.

Now if that picture had been taken obliquely, the parallel sides of the grid would appear to converge at the far side of the picture. This is the effect of perspective, for all the parallel lines would be running off towards the horizon and, if ex-

The place of photography in surveying

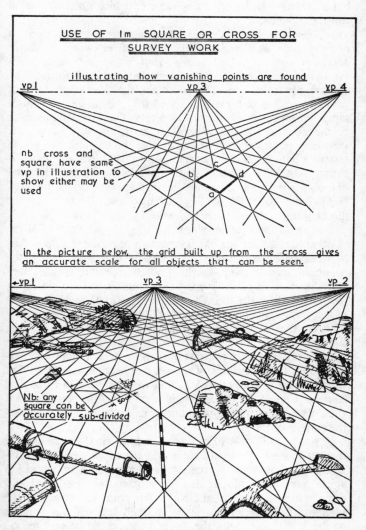

tended, would meet at points on the horizon line. The figure above will show that the squares do not appear as true squares but as tetragons, the opposing sides of which go to the same vanishing points.

So, given a print of an underwater area taken from an oblique angle, as it could well be from a diver swimming above and to one side or indeed from a low-flying aircraft, and assuming that there is one object within that picture known to be either a square or at least a rectangle, the sides of that object can be extended and it will be found that they meet at a vanishing point on the horizon line. Two sets of lines extending each of the sides of the square, both towards and away from the vanishing point are then drawn to give the beginning of a grid on the photograph. In the example (p 191), we will call the square ABCD. Next draw a diagonal to the square, choosing one that will give a convenient vanishing point on the horizon. Extend this diagonal to VP_3 and also away from it. In the illustration the diagonal chosen is AC.

To extend this system of squares over the whole area of the photograph, we use the old perspective principle that 'parallels between parallels are equal'. We know that lines joining the original square to their vanishing points represent parallel lines and therefore any line joining those parallels (and extending to VP_3, will also be the same length. Therefore, lines passing through points B and D, to VP_3 will also represent parallel lines, and so, diagonals of the two adjoining squares.

It is only necessary to draw lines through each point of intersection on the extensions of the sides of square ABCD, so that those lines go to VP_1 and VP_2 and we have formed three new squares. Repeat the process with other sets of lines to VP_3 and the area will be covered with tetragons representing squares on the surface of the site.

A good tip here is to buy plastics insert pockets as used in display folders. These are clear and closed on three sides. Insert your photograph in one of these and staple into position at the very edge. Mount this on to your drawing board and draw the lines with a suitable sharp point. In this way the photograph grid can be clearly marked out without damage to the print itself. Also, the marked plastic can be overlaid on the master drawing (assuming scale is satisfactory) and details can be pricked through.

The place of photography in surveying

The system explained here will allow you to position accurately, without correction, any objects that are on the same plane as the original square object which you used to establish the grid. Objects at different levels will require correction by the extensions of vertical lines to the original plane. Since the dimensions of the original square are known, we have a perfect scale that can be directly read in the vicinity of every object, for the tetragon adjacent to that object will have sides representing lengths equal to those of the master square. For even more precise location of an object appearing within the given tetragon, it is easy to draw in the appropriate diagonals and use the intersection to commence yet another and smaller square and so on.

No account has so far been taken of the effect of the distortion due to light passing from water to air, and as mentioned earlier, this should be corrected for, irrespective of whether the photograph was taken under water or from an aeroplane, for the error will be present in both cases. Unless, of course, the photographer was fortunate enough to have an Ivanoff correction lens on his underwater camera.

Only the simplest form of photogrammetry has been outlined here, but the science allows for much more complex problems. It is, for instance, quite feasible to work out heights, elevations, shadows, etc with complete accuracy. But this is too advanced to be dealt with here and those really concerned with this aspect of photogrammetry are advised to read Mr Williams' book.

PORTABLE SQUARES

So far it has been assumed that a known square exists within the framework of the picture, but, of course, this will not always be so unless steps are taken to ensure it. Metal grids have already been discussed, and one of these can be easily made to known dimensions, probably 1m for each side, subdivided into decimetre alternating black and white segments (p 194). If this is placed in the area to be photographed, it will be clearly seen in the ensuing print. But do ensure that it is level or, at least that it is level in one plane. Otherwise the

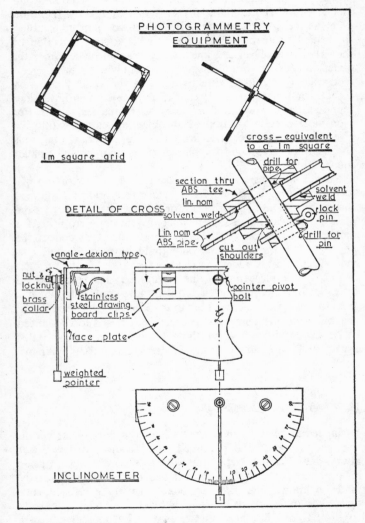

horizon obtained from it will not be true for the surrounding and perhaps otherwise level area.

The place of photography in surveying

CORRECTION OF SLOPING SQUARE (p 196)

It will help considerably to know the angle of inclination of the plane that is not horizontal and this can be recorded in the actual photograph at the time, if a device, such as the inclinometer already described, is fixed to the non-level side of the square. The free-hanging vertical plumb-bob will register against the protractor and will indicate the angle of the plane to the horizontal. As this has to show in a photograph, it will probably be better to paint alternate five-degree segments in black and white and ensure that the plumb line is painted a bright yellow. Every print will then clearly show the amount by which the plane deviates from horizontal.

When locating the vanishing points, the sides that are level may be extended in the usual way to locate VP_2 on the horizon line. But the sides that are sloping, if projected, would meet at a vanishing point on an entirely different plane and would therefore give a distorted pattern, so a correction must be made. To do this it is necessary to drop vertical lines from the raised corners of the square to the horizontal plane on which the lower level side of the square is resting. The plumb-bob on the inclinometer will clearly indicate the vertical and the drawn vertical should follow this. However, the lengths of these two verticals to the horizontal plane must be established. This is done by simple trigonometry using the sine rule and tables. Page 196 shows an exaggeratedly sloping square to help demonstrate the solution to the problem. We know that sides AB and CD are horizontal and their vanishing point can be established at VP_2. We have to locate points A_1 and B_1. Looking at the triangle AA_1D, we know the length of AD (the side of our square), we know angle A_1AD from the angle shown on the inclinometer, we know that AA_1D must be a right angle (90 degrees), and we can therefore calculate angle A_1DA. By using this information and the sine rule we can find the length of the vertical AA_1. From the measured point at A_1, draw a line to VP_2. Drop a vertical from B to B_1 on the line A_1VP_2. A_1B_1 now becomes a side of the corrected square. Extend DA_1 and CB_1 and they will meet on the original horizon at VP_1. We now have a tetragon A_1DCB_1 which represents a true level

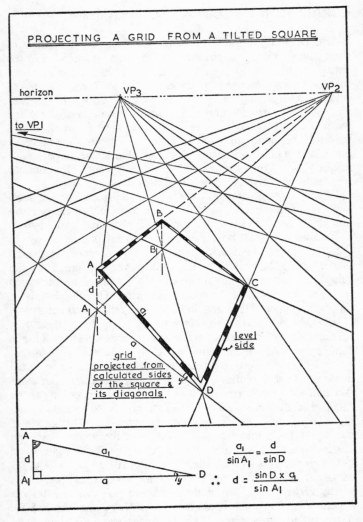

square from which we can extend the grid pattern over the whole area of the photograph. Obviously, we have not started from a perfect square under these conditions, but providing the square is not tilted too greatly in the first instance, the error will not be significant.

A development of a rigid framed square was described in Mr Williams' book, and was, I believe, developed by Miss Honor Frost. This is known as the 'cross' frame.

CROSS FRAME

If a cross is made in which both limbs are at right angles to each other and of equal length, then they can be regarded as diagonals of a square, the corners of which are at the extremities of the cross. If this is seen in a photograph, it is simple to join the 'corners' of the cross to make a square and project those lines to their vanishing points, as before.

A light-weight cross can be fabricated from alloy angle with two halves of one of the limbs hinged to the other limb and locked into an open position by two wing nuts and bolts, elastic, or even cords fixed to their extremities. Alternatively, the cross may be made from metal or plastics pipe of two different outer diameters. For convenience, when not in use the small diameter tube may be carried within the larger. When in use, it can be pushed through a hole drilled at right angles in the larger pipe at its mid-point. This area will require strengthening to prevent breakage. If made of plastics, two 'tees' can be cut lengthwise down their centre lines and then welded to the centre of the larger bore pipe. The smaller or cross bar is pushed through the branches of the 'tees' and the hole drilled in the main member. Correct location of the cross bar can be achieved by welding a 'stop' into place to prevent it from being pushed too far through and another 'stop' with a small bolt or spring clip to be slipped into place to prevent it accidentally being withdrawn (p 194).

Whatever the material of the cross, each limb should be painted in black and white segments to assist in scaling the grid. The length of the cross bars and the segments painted on them should be determined by the square of which they will act as diagonals. For instance, if you wish to use a 1m square for your grid, the lengths of the cross bars will be equal to the lengths of the diagonals of a square measuring 1m on its sides.

The sides of your 'square' will ideally be divided into decimetre segments, but in fact, those segments painted on to your cross will be more than a decimetre each. They will probably measure 14cm, for if you project 10cm measurements on a side of a square on to its diagonals, that is about what they will come to.

Of course, if the segments are there merely to offer a scale, any known length will do, but the scale will only be true in the immediate vicinity of the cross. On the other hand, if the segments reflect a fixed measurement on a true square, they can be projected all over the final grid with accuracy.

At the same time, because the grid is based on a relatively small module (1m) which can be accurately sub-divided in halves by the mere drawing of a diagonal in any specific tetragon, a half-metre scale can be obtained without recourse to segments painted on the cross and it may be considered that those would be sufficiently small sub-divisions. To some extent, this is a matter of personal preference.

We have been discussing the use of photography as a survey aid, but it may also be remembered that it is a most important asset for detail recording. Here it is just a question of knowing the techniques of underwater camera work to be able to obtain, consistently, the required standard of picture.

Every new find should be photographed in situ together with something to provide a scale. Even if a rule is not available, the use of a diver's knife, for example, will often prove adequate, for this can later be measured exactly. Ideally, I believe that every expediton should have one member who is really competent to carry out underwater photography, and this, and its associated tasks of development and printing, should be his full-time work and responsibility. I also believe that cine film should be shot on site if at all possible. In this way, new techniques can be filmed and used for training purposes at a later date; often also the viewing of such a film will indicate areas where improvements of technique can be effected.

But it is not very satisfactory for the one man to try to do both kinds of photography. Effective cine coverage requires a very great deal of concentration and awareness of the activities that are taking place, and also tends to require the photo-

Page 199 (above) Fifteen different types of amphora, all found in an area 150m x 300m in the west harbour area of Nora, Sardinia; *(below)* a view of the quay/quarry area of the Roman port at Tharros. Remains of buildings can be seen leading down into the water, which covers the quay area to a depth of about 30cm. The port area was assumed to be some 1,500m north of this spot and adjacent to the actual town. The survey established that, in fact, rock had been quarried here for the building of the town and the area levelled to make a quay

Page 200 (*above*) Two bronze cannon raised from the *Amsterdam*. The letters VOC (Verenigde Oestindisch Compagnie) and the date 1748 identified the vessel. Also shown is a cannon ball with brass 'shot gauge' to check the size of shot; (*below*) noting bearings and measurements and checking rough bearings by compass, which is being held squarely in front of the diver to ensure a correct line

The place of photography in surveying

grapher to be at a distance from the action and not right close up. On the other hand a great deal of the value of still photography stems from the ability to take close-ups, so immediately we have a conflict of interests.

In certain very ideal conditions of calm seas and clear waters it will be possible to take photographs from the surface through a glass-bottom pail or even boat. Very effective cover can be obtained in this fashion, but do not fall into the trap of incorrectly focusing your camera. Just because you and it are above water, does not mean that a correction for the image distortion at the air-to-water interface is not necessary—it is.

Also the correct exposure will be a little difficult to establish due to the amount of light about you. It helps to shield your head and camera under a cloth, so that the interior of the pail or boat (as the case may be) is relatively dark. In this way, just as with the old plate cameras, the photographer will obtain a far clearer view of the underwater subject and will be able more correctly to assess his exposure.

I think we have covered all that we reasonably can without going into the technical details of photography and equipment which I do not feel to be within the scope of this book.

11

UNDERWATER EXCAVATION

THE FIRST thing that should be stressed here is the absolutely essential requirement of obtaining permission before raising anything from the seabed if one is abroad. Even in one's home waters this is advisable if not essential—see the notes on law and salvage in Chapter 3.

If, by good chance, you have the opportunity of diving abroad, particularly in the Mediterranean, whether on an expedition or as a private party, your diving activities will be very quickly be brought to the notice of the authorities. Frequent and unannounced visits will be made to your camp site by very polite and courteous police or other officials. If it is seen that you have been raising artifacts from the seabed, however insignificant they may seem to be, you may have some difficult questions to answer. In some countries, they may not even wait for the answers before hauling you off to the local jail—and this is quite serious; I am not exaggerating.

The activities of divers throughout the Mediterranean have led to a depletion of archaeological finds from innumerable wrecks. Just as on land where tomb robbing has become almost a way of life for the local peasants, so the seabed is being scoured. Wrecks so found, are being plundered with little or no regard to the total archaeological value. As a result, divers of any race or creed are viewed with the same suspicion.

In all of my expeditions to Sardinia, I have ensured that we have obtained permission in writing to dive on the sites upon which we have wished to worked. Even so, we are frequently

Underwater excavation

visited by the police and, almost as frequently and at very irregular times, we have had police helicopters circle us overhead. Their crews could be seen inspecting us and our camp site with field glasses and I doubt whether very much escaped their search. So for your own sakes and those of other divers, go about the job in the correct fashion.

Having delivered that little homily, let us get down to actual site work! Excavation is possibly rather a strong word here, for in many cases, we shall be concerned with the random search and collection of artifacts that have been scattered over a wide area. More of a problem in finding than excavation, for once found, it is only a question of picking them up!

Starting with the tools that every diver has with him—a diving knife, flippers, and a pair of hands—what can we really say about their use? To start with, recognition is not always easy and the use of the knife for tapping has been mentioned; a change of sound can indicate a substance of interest. A knife can be used for scraping weed and encrustation from a possible find and it will come in handy for digging something free of the sand or silt that may be partly covering it. Even with a light and comparatively delicate tool such as a knife, do treat every possible artifact with care.

A partially buried pot, amphora, or even a metal object may be very much weakened at the level of the sand. This could be due to the scouring action that takes place there, or possibly to the effect of surge over many years. Excess of strength in trying to pull it out will merely cause a break at that point and you will have at best a broken sherd, rather than a complete object.

WASHING SAND AWAY

On first sighting an object, try to resist the temptation to prod it with your knife, try a more subtle means of freeing it first. Use your hand to sweep the sand away from its vicinity. If the sand is very hard, gently loosen it some way away from the object with a knife. Then pick the pieces up and create a current of water over the object with your hand. This is best done by sweeping the cupped hand over the object quite forcefully and bringing the hand back to the 'up-stream' side either in a wide circle or if space does not allow that, by 'feathering' your hand and slowly bringing it back, just as a good oarsman feathers his oar on the back stroke. Aimless flapping of the hand over a sandy area will merely cause the sand and silt to rise and cloud the water, but the method suggested will result in a current of water sweeping across, loosening and taking the silt etc with it. Your visibility will remain good and surprisingly large quantities of sand will be moved, gradually working the artifact clear.

If the sand is very soft, take off a flipper and use this instead of your hand, but in the same way. Frequently these methods will completely loosen the find, allowing it to be picked up freely. Incidentally, it pays to inspect the immediate vicinity carefully first—to locate any sea urchins on to which you might sweep your hand, with unfortunate results. If it is anticipated that a fair amount of this sort of searching is likely to be called for, a paddle such as a table tennis bat or toy plastic paddle can be taken down and used to wash away the sand.

Try not to remove any more encrustation than may be absolutely necessary to free the object, because it may be covering a handle or narrow projection on the objects it surrounds. Careless treatment on the bottom might result in these being knocked off, whereas on shore, careful treatment would save it.

Still using hands, flippers, or bats, endeavour to wash sand and silt out of cracks and ledges that may appear in rock structures. Relatively small heavy articles especially coins, tend to work their way down through sand and rest undisturbed by surge and swell in such places. Two of our most important pottery finds, a fifth-sixth century BC jug and an almost perfect shallow bowl, probably of about the first century BC were found lodged into cracks in rock walls. A long thin probe of spring steel (stainless for long life) can be excellent for scratching into narrow cracks.

ENTRENCHING TOOLS

These can be quite useful, combining as they do both shovel and pickaxe. But they are not easy to use underwater, for there is so little purchase for the diver. The tendency is for the diver

Underwater excavation

to move rather than the tool! In soft sand they can be used for fanning and also for scraping holes, but the limitation here is the ability of the sand to return to the hole quicker than the diver can dig it out. But, since these tools can be used with a pulling action, they are useful in situations where it would probably be impossible to use a spade, and in confined areas where it is not possible to get heavy equipment, they can be very useful. We always have a couple along with us.

CROWBARS AND JACKS

The time will come when the team is faced with the necessity to move large boulders which are too heavy for a 'straight' lift. Crowbars can well be used, but take care when inserting the point; it could be that there is something valuable just where it is placed. Frequently, heavy objects resist movement, not so much because of their own weight but due to suction of the surrounding sand or mud. So take care, as soon as any movement is felt, ease off the pressure otherwise the boulder may move too rapidly for control and the crowbar could slip out and cause injury.

Blocks too heavy to be moved even with a crowbar can sometimes be moved with a small hydraulic jack. After inserting one of these between the object to be moved and bedrock or a nearby heavier block, a few strokes on the handle will break the object free from mud or whatever is holding it and a crowbar will complete the task. Obviously care must be taken to ensure that any movement resulting from this action will not cause damage to other finds when the block moves.

Heavy items such as cannon or anchors will respond to this treatment, particularly when partially covered with coral or general encrustation. But it is advisable to use this form of lifting only after every effort has been made to free the object by gentler methods.

HAMMERS AND CHISELS

A short-handled mason's hammer and a cold chisel is an excellent combination. It is virtually impossible to swing a long-handled tool, especially if one is working in a swell. One of the funniest scenes I have seen under water, was a diver trying to break off samples of rock for identification purposes. He was using an ordinary pickaxe and in a strong surge caused by the wave action over his head, was swinging hopefully at his target. About one swing in three hit the target and with very little effect. A short 7lb mason's hammer and a cold chisel would have done the job in moments. When using these tools, try to locate the edge of the chisel on existing fault lines in the concretion. These can usually be seen, sometimes adjoining the body of the find. Use very slow and firm blows, and stop frequently to check your progress and to ensure that damage is not being done to something of importance. As in all things underwater, 'hasten slowly' is a good slogan.

CURRENT BARRIERS (p 207)

It is nice to be able to get work done without personal effort and this is sometimes possible if one can harness the tide or current. If the site happens to be in an area where tide or current runs fairly fast and where, in course of time, the wreck has become silted over, it is sometimes possible to harness that tide or current. Just as water sweeping between two boulders will wash the passage between them clear of sand, so an artificial barrier sited in a strategic place will deflect the current and can cause a scouring action where you want it.

Assume for the moment that it is a wreck mound or something similar that we are working on. Starting at the point farthest upstream, a stake or scaffold pole is driven very firmly into the bottom about 1m to one side of the mound. On shore we will have prepared a sheet or two of galvanised iron by drilling two pairs of holes, one set above the other at each end. Taking the sheet down, one end is rested against the upstream side of the post and secured to the post with strong loops of wire, not binding it too tightly. Pieces of ribbon or cloth about 1m long are tied to both top and bottom wires.

Now the free end is swung around upstream of the secured end (it may take several divers to do this in a strong current), until the ribbons are being deflected by the current towards the mound. When they are hitting it at the right point, insert another scaffold post behind the free end of the sheet, firmly

Underwater excavation

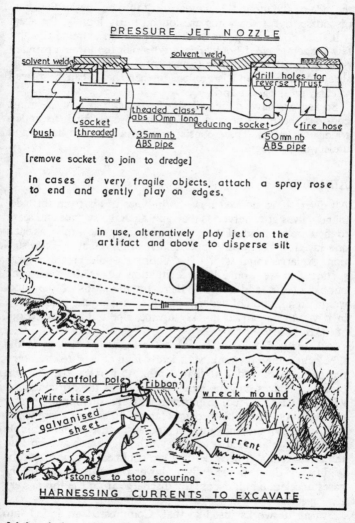

driving it into the seabed again and secure the sheet to it as with the opposite end.

Over a period of time the deflected current will wash away the side of the mound. It may also try to scour under the sheet of course, but if the wires are looped fairly loosely, the sheet will slide down the posts as this happens. In extreme cases, a few moderate sized boulders placed along the lower upstream edge of the sheet will help to prevent the scouring beneath the sheet. As the mound is eroded, further sheets can be placed to carry on the good work.

So far we have discussed simple tools, and now we must consider the more powerful tools that are essential to underwater excavation. These consist of airlifts, pressure hoses, and underwater dredges.

AIRLIFTS

All divers have noticed how the bubbles of air from their demand valves grow larger as they approach the surface and seem to increase in speed. If these bubbles were inside a tube, as each one raced for the open upper end it would create a little suction. A large number of these bubbles would create a large suction and we would have a situation where, between each bubble of air, and trapped there by the walls of the tube, would be a section of water. The water and anything contained in it would be carried quickly up the pipe until it spilled out of the open top.

All this is due to the fact that the air at the bottom of the tube is at a high pressure, sufficiently high at least to overcome the pressure of water at that depth. As it rises, the ambient pressure reduces and the bubble expands as the air contained within it tries to equalise pressure with the surrounding water pressure. As the bubble becomes larger it displaces more water and becomes more buoyant. On land we have a similar example in the very high factory chimney. The hot gases are fed into the lower end and because they are confined to a narrow tube leading up through successively lower pressure air layers, the gases accelerate and eventually discharge from the top at very high speeds.

An airlift works just like that. Compressed air is fed into the lower end and water is sucked up; in the process anything that is light enough to be moved by the resultant current of

Underwater excavation

water is carried with it. A simple device and excellent for our work. The power of an airlift should not be underestimated: what is 'light enough to be moved' could be a very substantial boulder, and providing it will pass freely through the bore of the pipe, it is quite conceivable that it will be lifted up and ejected at the top. So it is as well to be careful in the location of the top, otherwise the diver may get back more than he bargained for!

In its simplest form, an airlift requires a piece of pipe and an aqualung. Take a tube from the latter, 'crack' open the valve and let air into the open bottom end of the pipe and you have an airlift that will move sand and light objects. Obviously, it would not be as efficient as those we shall discuss and it is not too good for the aqualung. The comparatively rapid exhausting of the high-pressure air will cause excessive cooling of the bottle and condensation will take place, leading to rusting. But for an emergency lift, it can be used.

An airlift can vary from perhaps 50mm bore to 250mm—conceivably larger, although the larger size mentioned would be too much for a non-professional diver to handle and even with skilled operators, would require powerful lifting equipment and air compressors. Its length can be from only 3m to 20m, but it is relatively inefficient in lengths of less than 10m. The reason for this is that air doubles its volume for roughly every 10m it rises. Conversely, air has to rise 10m to double its volume and so to begin to generate the lift capability that is latent in the device. For this reason, although it can be used in shallow waters, it is frequently more satisfactory to use the water dredge, which we shall describe later on, in water less than 5–6m in depth.

The suction pipe should be within, say, 30 degrees of the vertical. With any lesser angle, the solids that are being carried up the pipe will tend to slow down due to surface friction against the lower inner pipe wall. Also, the pipe will almost invariably be in sections joined by couplings of some sort, which will tend to trap the mud and sand if the angle is too small, which will lead to blockages. On the other hand, before making it vertical it is as well to remember again, that what goes up must come down—and a cannon ball toppled out of the top could make a nasty dent in the operator! A 30 degree slope down stream or tide is ideal as then heavy articles that may fall back, will be well clear and at the same time, general mud and sediment are washed away from the work area.

The airlift pipe will require anchoring to hold it down, yet buoyancy to hold it upright. An apparent contradiction, but nevertheless quite true. When not in operation, it is undesirable to have it lying across the site where it could possibly cause damage and would certainly waste time in getting back into operation, so a buoyancy device should be rigged near the top. It will be necessary to anchor it down, not only to locate it when floating vertically and not in use, but even more so to locate it when the lift is in operation. The forces of rising air and water will both add buoyancy and a lifting force which must be counteracted. The type and size of anchorage depends upon the size of the lift, the permanency of positioning, and the size of the site.

The airlift used at Mahdia by the Club for Underwater Studies of Tunisia was relatively lightweight, yet they anchored it with lines round two winches weighted to the bottom. By unwinding one and winding in the other line, they were able to move the airlift over a fairly wide area. On the Grand Congloue site, the lift was suspended from a derrick mounted on rocks at the water's edge. This could, of course, be swung round to enable the lift to be used over a wide area. The design of anchor points requires some thought, for winches and derricks are not so easily available.

At the same it is not feasible to use really heavy weights, because they will almost certainly have to be moved to allow full coverage of the site, unless you are working in very deep water. On a deep site, it is possible to space the anchorage points well apart and use long fastening lines, secured well up the lift. In this way, by adjustment of the lengths, the pipe may be moved and still remain stable. In shallow water, the lines would be stretched too flat to provide a good anchorage if the anchor weights were too widely spaced, hence the need to move from time to time (p 212).

It is wisest to use three anchor lines for positive location and these are best attached to the lift at a point fairly near the centre of gravity (allowing for the buoyancy) of the pipe system. In this way, it will be easier to swing the lower end

Underwater excavation

around to wherever it is required. If fastened too high up, there will be a 'counter balance' or pendulum effect at the lower end to be overcome by the diver. If fastened too low, the buoyancy device at the top will be trying to resist any movement. In either case, an unnecessary waste of physical effort is called for.

If a large-bore lift is being used, the lower end can be made of a reinforced flexible pipe, which also helps manoeuvrability. But I do recommend flexible joints at various points up the actual lift itself. The straighter it is, the more efficient it will be and flexible joints, by their nature, will allow bending to some degree. If one is using plastics pipe of the rainwater, soil, or pressure type, which incidentally come in a handy range of diameters—from 25mm to 300mm (even up to 400mm in some cases)—and for which integral couplings are provided, the job is made much easier. As remarked before, plastics have many advantages, but in the case of airlifts we have the additional one of low specific gravity, so less additional buoyancy is needed; also they are lighter to swing around and have a smoother bore which will improve efficiency. Try to obtain pipe which is complete with a ring-seal coupling, but not an 'O' ring, if possible. What is known as a 'multi-ridge' seal will provide a better seal and will also resist being pulled apart. The old 'O' ring system allowed the pipes to be separated as easily as they were jointed and so would not be ideal for this job. Even with the better seal, if the lift is to be in use for any appreciable length of time, I would recommend 'splinting' the connection between two pipes.

Two or three strips of wood laid lengthwise to the pipe and across the joint can be lashed tightly with cords and will do, but a similar number of thin strips of copper, again laid over the joint and along the pipe will make a better job. Above and below the joint, place a 'Jubilee' type clip over the end of the strips, tighten up and then bend the ends back over the clips. A neat and workmanlike job, which will stand considerable strain, is the result (p 212).

As we have said, the lift will be quite capable of sucking large pieces or objects into its maw and these could cause a blockage. It can be a very tedious job trying to clear it. However, such blockages can be prevented by artificially restricting

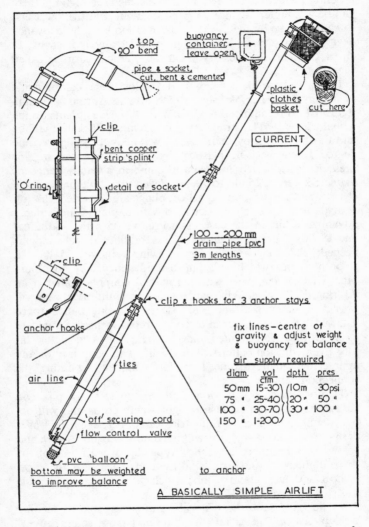

the open end of the pipe. In the case of small pipes, what is known as a 'balloon', in other words the wire or plastics basket

Underwater excavation

affair placed at the top of soil pipes to stop birds from falling down, will do very well. On larger pipes a metal ring can be made which will clip over the end of the pipe and so restrict the opening. Anything passing through the ring must be smaller than the bore of the pipe and, therefore, is unlikely to block it.

The top of the pipe can be treated in two ways. If depth is not too great and the lift pipe is long enough, let it rise out of the water and discharge over a weir and into a net or sieve arrangement. This will trap even the smallest article and only needs a watcher to throw out the pebbles, weed and other odds and ends. The sand, silt, and mud will be washed through, leaving the 'goodies' in the sieve. Again ensure that the outlet is downstream of the site to prevent spoiling the visibility by the disgorged sediment. Also, and very important, do not take the end any higher out of the water than absolutely necessary. Once water is lifted within the pipe to above the sea level, it is a dead weight and the greater the length of pipe above water, the greater the weight of water being lifted, and the less efficient the lift will become, just as a pump loses power and output as it pushes against an increasing 'head' of water. A good arrangement is to fix the top of the pipe to the side of a large inflated truck inner tube. If the open bottom of this is covered with a fine net it will make an excellent trap which will not necessitate lifting water more than about 20–30cm above the surface. A suitable 45 or 90 degree bend at the end of the pipe will guide the outfall into the net.

The alternative method, almost unavoidable in really deep water, is to attach a net or basket at the head of the lift, again after fitting a bend, so that the outfall enters the bag and once again, any items of value are trapped while mud, etc, flows through and are washed away downstream.

In fact, both of the traps are really only safety devices, for it is far better that the operator should not just ram the open end of the lift into the bottom and suck away great gobs, hoping the trap will catch everything. In the first place, all sorts of unnecessary rubbish will be sucked up, adding to the risk of blockages and certainly filling the trap with a heavier load of rocks, etc than it could probably carry. In the second place, an airlift can be very powerful and therefore destructive—any delicate article coming within range of its suction will be whisked up and probably smashed to pieces. Far better to keep the suction end well above the work area and gently loosen the sand or mud as you work and let the spoil be swept up the pipe. Odd pieces of stone and junk can be examined and thrown to one side if of no interest. Worthwhile finds should be noted, with details of position, attitude, and description, before being picked up and placed in a handy basket.

In this way, we can, at comparative leisure, examine the archaeological significance of each find. Delicate items such as leather, rope, and basketwork can be recovered intact.

AIRLIFT REQUIREMENTS

I have tried to explain how the actual lift works, so let us now consider the motive power needed. The actual amount of air that is required is dependent primarily upon the bore of the pipe. The following figures will be a reasonable guide.

AIR REQUIREMENTS

pipe internal diameter	supply hose internal diameter	air requirements cubic metre/minute	output volume litres/minute
50mm (2in)	12mm ($\frac{1}{2}$in)	0.28–0.84 (10–30cfm)	90–227 (20–50gpm)
75mm (3in)	12mm ($\frac{1}{2}$in)	0.28–1.12 (10–40cfm)	182–362 (40–80gpm)
100mm (4in)	20mm ($\frac{3}{4}$in)	0.81–3.11 (30–110cfm)	540–1080 (120–240gpm)
150mm (6in)	30mm ($1\frac{1}{4}$in)	1.40–5.6 (50–220cfm)	900–1800 (200–400gpm)
250mm (10in)	50mm (2in)	4.2–12.0 (150–450cfm)	3,000–4,000gpm (700–800gpm)
300mm (12in)	50mm (2in)	5.60–15.6 (200–550cfm)	4,000–4,600gpm (900–1,000gpm)

It may be noticed that there is no reference to pressure; this

Underwater excavation

is because pressure is relatively unimportant. So long as the air pressure is even a little greater than the water pressure at the depth at which we are working, air will come out of the hose, and that is really all it has to do. However, volume is another matter. Plenty of volume is required and this, in turn, means quite large compressors, in fact very large in the case of very big airlifts. Bottle-charging units are no good, they produce little air at high pressure—and we want lots of air at a relatively low pressure, rarely more than 100psi. You may also notice that the output volume does not rise proportionately to the increased air volume. At the lower end of the scale the output is approximately four times the volume of air put in, but at the top end this falls to roughly double. This is not surprising, for one could imagine that if so much air was forced into the pipe that it became full, it would not carry water at all! Indeed this can almost become the case, so a means of controlling the air intake at the lower end of the lift is needed.

This generally comprises a turn-cock or lever valve, which the diver can operate while he is working and so control the volume and speed of water up the pipe. A word of warning here: if you should use a lever type of cock, make sure that it is in a position where it cannot be accidentally knocked on, or where you can put a safety catch to lock it in the 'off' position. A commercial diver friend of mine told me of a case where a diver had left his airlift (a big 250mm one) with the lever cock in the off position. Possibly due to surge or rolling, the lever was accidentally returned to the open position. Even though the main pipe was lying slightly off vertical, it quietly and rapidly dug itself straight down into the sand until eventually the airline was stretched to breaking point. It was not until the airline actually broke that the digging stopped, by which time the main pipe of the airlift was out of sight under the bottom. It is still there today! The power of an airlift has to be seen to be believed.

The drawing on p 212 will, I hope, help you to make your own airlift. I have shown a 4in drainpipe (actually 110mm) as this is the size most likely to be suitable for normal activities and is easily obtained from any builders' merchant, together with all the fittings.

The only comments needed to amplify the drawing are that

the air inlet should be as close to the bottom as is reasonably possible, probably around 30cm as a maximum. It may be found convenient to incorporate some form of handle at the lower end for the operator and it may be found necessary to weight the lower end slightly—trial and error will decide this. To prevent large objects being sucked up, take a 'socket piece' —these can be purchased separately—and solvent-weld a couple of strips of plastics material across the narrow end, or thread a couple of strong piano wires across the open end. This can then be slipped on to the bottom of the lift and yet can be removed easily if a need arises. To give some idea of the weight of an airlift made from plastics materials I have shown below the approximate weights for various sizes of pipe:

Material—UPVC (specific gravity 1.37)
50mm (2in) = 8kg per 10m
75mm (3in) = 13kg per 10m
110mm (4in) = 16kg per 10m
150mm (6in) = 27kg per 10m

To this must be added the weight of the pressure hose, fittings, etc to arrive at the buoyancy required. But as the specific gravity of the plastics material is so low, relatively little extra buoyancy will be needed. This may again be a case of trial and error and will in any case depend upon what is used. But to help you, I detail below some displacement values for spheres and cylinders, which will help if you are using inflatable balls or tubes to support the lift pipe.

Seawater Displacement

Spheres		Cylinders	
Diameter	Displacement	Diameter × Length	Displacement
150mm (6in) =	1.8kg (4lb)	150mm × 250mm =	5.4kg (12lb)
250mm (10in) =	9.0kg (20lb)	150mm × 500mm =	10.8kg (24lb)
300mm (12in) =	15.3kg (34lb)	250mm × 250mm =	13.5kg (30lb)
450mm (18in) =	45.4kg (100lb)	250mm × 500mm =	27.0kg (60lb)

If non-inflatable buoyancy is used, such as polystyrene, ensure that it is strongly wrapped in a canvas or similar cover and then lashed by broad bands to the pipe. Foam plastics of this type are notoriously weak and will break or crumble easily if not protected.

The air feed tube should be strong enough to stand the highest pressure the compressor can provide. When the valve is open, there will not be any strain on the tube, but when the valve is closed, partially or completely, a back pressure will be set up and the pipe could burst. Nothing but inconvenience will be caused should this happen, as the surrounding water will absorb the burst, but a split hose will require a repair and so time will be wasted. I recommend that the hose be secured to the main airlift tube by light fastening so that it does not drift all over the place.

HIGH PRESSURE WATER JET

This is, as the name implies, merely a hose through which water is pumped so that it comes out of a jet nozzle at relatively high velocity. It is a very powerful tool for removing mud, sand, and light shingle. Fire hose is ideal, for this rolls up flat, and so takes up litle space, yet will stand all the pressures likely to be required. It can be obtained in long lengths and possibly, if asked for in the right way, from your local fire station. They replace hoses at fairly regular intervals as a safety measure. The old hoses are scrap and generally they are happy to part with them, sans fittings!

Fittings may possibly be expensive, so make up your own from ABS pipe. Try to get 50mm (2in hose) and you will find that you can fix a short length of pipe into it with a spring clip and add to this reducers or composite unions as shown on p 207 to complete the nozzle assembly.

I would suggest that holes should be drilled around the nozzle head, just upstream of the first restriction. Back pressure at the restriction will force water back through the holes and build up a reverse thrust, on the principle used on large 'recoil-less' artillery. Without this arrangement a diver has great difficulty in controlling the jet and can easily end up disappearing backwards at a rapid rate of knots! Alternatively it is possible to purchase specially made 'Zetterstrom' or

'Galeazzi' nozzles but these are likely to be difficult to obtain and rather costly.

It will almost certainly be necessary for the diver using the jet to hold on to a hand-hold to keep in station. However, by increasing the jet size the velocity of the water exiting from it and so the back thrust can be reduced. It is therefore a good idea to have a variable-orifice nozzle or at least (as is illustrated) one which can be altered by removing the various reducing stages.

The motive power for this equipment can be any water pump, but one should be chosen that is designed to handle salt water, otherwise corrosion becomes a big problem. Even with such a pump it is as well to wash out after each period of use by pumping fresh water through it. It will not matter whether it is a centrifugal, piston, or diaphragm pump, because it will only be handling water and no solids or suspended mud, etc. Although a pump with as low an output as 90 litres/min (20gpm) will be adequate for the jet, it is best to choose a larger capacity pump, for it will then be suitable for the underwater dredge to be described later.

Most pumps are rated by suction and delivery head as far as output is concerned, in other words, a pump may deliver 100 litres/min to a height of 3m, but it would only deliver 50 litres/min to a height of, say, 6m. Similarly the height to which the pump has to suck the water up will determine its output. As we shall be pumping water down from the boat to the bottom of the sea, we need not concern ourselves about the 'head' of the water. Nor do we have to concern ourselves about overcoming the pressure of water at the depth at which we are working.

This has no bearing on the matter, for as you probably know, the water pressure at any particular depth is really the weight of the column of water above that point, plus one atmosphere. As we shall be taking our own 'column of water' down with us, wrapped up in a hose, this, plus the same one atmosphere of pressure at the surface, will equalise the pressure at the nozzle, even without the pump working. Therefore a pump capable of delivering 100 litres/min at zero head will also deliver 100 litres/min at 10m depth.

On the other hand, the suction head has an important part

Underwater excavation

to play, especially with low-output pumps. It is therefore essential that the pump be situated as near to water level as possible, preferably in the bottom of the boat. It will not matter if the intake hose has to rise over the gunwale or inflated hull before dropping to the inlet port of the pump, as the rise and fall cancel out. Nor does it matter if the suction filter is dangling well down in the water. What is critical is the height of the intake manifold above the water level, so place the pump as low as possible. I have used a Forcam Engineering Ltd 450 litre per minute per pump, with excellent results, in fact far too great an output for anything smaller than a 40mm (1⅝in) nozzle, but quite adequate for the dredge.

Most portable pumps are weatherproofed, but they still do not take kindly to salt water, so protect them as much as possible from spray, etc. Also, especially in inflatables, do be careful how near the exhaust port is to the rubber sides, for the heat output is pretty high and can do a lot of damage to a rubber hull—not to mention a swimmer's bare legs! See, too, that the exhaust is pointing down wind from the boatmen. The combination of a nice swell and a not-nice smell can daunt the hardiest sailor!

Ensure that the hose is well rolled up on the boat with nozzle attached before the work starts. On getting to site, hand the rolled-up coil of hose to a diver already in the water. If he swims down backwards, ie facing the boat, it will be possible for him to pay the hose out as he sinks, without kinking it. It does not matter all that much if it does kink, because the water will soon straighten it as the pump forces it through, but it looks nice and professional to get it unrolled without it resembling a lot of knitting.

On the given signal from the snorkel cover swimmer, the boatman starts up the pump and the diver is usually propelled backwards until the hose is stretched out straight before he can get organised. Having orientated himself, the diver will try to swim back to the desired work area and will probably get nowhere against the pressure of the jet. The crafty one will wind the hose over his shoulder so that the nozzle points down across his chest, and then aim himself towards his objective, when the jet will propel him along without any effort.

In the work area, the method of use will depend upon the

work to be done. If it is just a question of moving vast quantities of mud, sand, or silt, then the operator will anchor himself or the hose nozzle, if necessary by driving a stake into the bottom for a long job, and play the jet on to the area to be cleared. He should start at the top of the mound and wash it away, rather than starting at the base where he will just dig a hole into it and will stir up vast amounts of muck to the detriment of visibility. It generally pays to play the jet on to the objective in short sharp bursts, in between raising the jet above the work area and directing it horizontally. Say, two seconds down and three seconds horizontally. In this way the disturbed mud which swirls about, will be washed away from the area in the direct water current from the nozzle, enabling the visibility to be kept quite good in the work area. Obviously, if there is tide or current, try to work facing downstream.

The power from even a quite small pump will move relatively large boulders, and small pebbles, etc will be swept away with considerable force. The noise of the stones rattling along, will sound loud to the operator. Accidental direction of the jet on to a pot or delicate article is quite likely to smash it to pieces before you can stop, so care must be taken.

If working in an area suspected to contain breakable objects, it is important to reduce the nozzle velocity to controllable limits. If this can be done by throttling back the pump, so much the better, but most portable pumps work flat out with a governor acting as a control and cannot be adjusted. In such cases, open up the nozzle to a wider aperture, if using an adjustable: if not remove one of the reducers as mentioned earlier. Then with the jet as light as possible, nibble away at the edge of the area. Use two divers, one holding the jet and the other close to the point of impact of the jet current, keeping a lookout for anything of interest. Once again alternately play the jet on to the target and then horizontally, gently working away at the area until all has been uncovered.

Summing up, the pressure jet is a useful piece of equipment which is easily handled, very effective in use and not too costly to obtain. But as with so many pieces of equipment it can be dangerous, so use it with care.

UNDERWATER DREDGE (p 222)

It was mentioned earlier that an airlift is most effective in depths greater than approximately 10m. At lesser depths than this, it gradually becomes less effective, because the difference in pressures is relatively small. Given a water pump, we can adapt an idea used by Canadian prospectors in their search for gold in the fast-running streams and rivers of Canada.

Many readers may remember chemistry and physics lessons at school and will recall that liquids were pumped by the use of a venturi pump. This consisted of a venturi tube which was connected to a water tap. At the point of the venturi a branch pipe was connected. When the tap was turned on, the rush of water through the venturi caused a suction or syphonage at the mouth of the branch pipe and the liquid was drawn up the pipe and out through the venturi. The dredge works in precisely the same way.

It consists of a tube, like that for the airlift, but with a 30 degree bend positioned at the foot. At the apex of the outside of the bend the water jet from the pump is attached. This jet is aimed straight up the centreline of the main pipe and forces the water in the pipe along, creating a suction at the mouth of the main pipe. This suction can be quite powerful and will move very considerable quantities of matter. The height of the lift possible will depend upon the bore of the pipe and the output of the pump. A large 900 litre per minute (200gpm) pump and a 150mm (6in) pipe will lift as high as 20m, and at lower levels can move as much as 7.5 cu m (10cu yd) of loose gravel mud, and sand per hour.

Using a 110mm (4in) pipe and a 450 litres/min pump, we found we could lift about 5m with a 10m long pipe, but not too effectively. However, there is no reason why the dredge pipe should go up—it can equally effectively be aimed to one side of the site and the spoil deposited quite clear of the work area. The same pump and bore of pipe carried our spoil a good 25m away and downstream of the work area. For all practical purposes this was as good as taking it upwards and there was no risk of large objects being cascaded back on to the hapless diver. Once again a sieve can be rigged at the

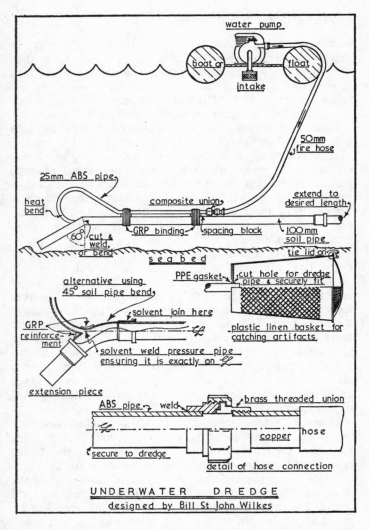

outfall end, although with proper care at the suction nozzle by the attendant divers, objects will be found before they are

Underwater excavation

sucked up. The tube can be quite light in weight, for preference made of plastics, and re-positioning is simply a matter of picking it up and swimming it to a new position. To alter direction, merely pick up the outfall end and swim that round until the suction end is facing in the correct direction. No complicated anchorages or buoyancy are required and it is all very simple.

The illustration opposite will give an idea of the method of making a dredge. For ease of handling, the flexible pressure hose is connected to a small-bore tube that is laid parallel to the main pipe at the suction end, until it is about 450mm (18in) from the entry point. It is there swung round in a gentle radius and connected permanently to the intake nozzle. It is important not to use sharp bends anywhere on the delivery line for they would reduce the pressure at the jet by slowing down the flow through the line. At the free end of the attached pipe, fix what is known as a composite union (also obtainable from a builders'/plumbers' merchant); this will enable you to connect the same hose and nozzle as are used on the pressure jet.

There is nothing to stop you connecting the hose directly to the input point (except that the hose might kink and constrict the flow) and tying it to the main pipe system. However, I would recommend going to the extra trouble of making a rigid connection as illustrated. Incidentally, this becomes a handy loop through which the diver can insert his arm while working the dredge. Left arm through loop and right hand free to sort out the bits and pieces which are uncovered.

The speed of the water entering the bend of the main pipe needs to be as high as possible, so keep the intake pipe down to as little as 25mm (1in) for the smaller dredges. It would help if a proper venturi restriction could be built into the bend of the suction pipe, but this is probably beyond most amateur workers, so we cannot achieve maximum efficiency. It is even more important, therefore, to ensure that the inlet is accurately placed in relation to the centre line of the main pipe. I list below some comparative sizes of inlet pressure nozzles for different size dredges and maximum pump outputs, required to operate them efficiently:

Suction Dredges

Suction Pipe Bore	Pressure Inlet Bore	Recommended Minimum Pump Output
50mm (2in)	25mm (1in)	235 litres/min (55gpm)
75mm (3in)	25mm (1in)	450 litres/min (100gpm)
100mm (4in)	38mm (1½in)	550 litres/min (125gpm)
150mm (6in)	38mm (1½in)	1,350 litres/min (300gpm)
200mm (8in)	75mm (3in)	2,250 litres/min (500gpm)

In relatively calm river or lake waters, it is possible to mount the pump and motor on inflated inner tubes with the outlet end of the dredge discharging into a sieve on another, attached, inner tube. In this way a very lightweight and portable unit can be devised for the divers to use anywhere and even to tow above them as they work the bottom. Such a device is being used by divers in the United States to recover golf balls (under contract with the club, I hasten to say) on courses with water hazards. I believe it is a thriving business too.

You will note that the pump itself is still only handling clean seawater and is not called upon to pump solids or abrasive matter. So a centrifugal pump, most of which are claimed to handle solids up to half the size of the inlet port (an unnecessary requirement with our equipment), is likely to be the easiest and cheapest to procure. The pump and engine should be obtained as a combined unit, of course.

We have now covered the types of equipment which are likely to be of the most value in underwater excavations. Each has benefits in certain directions, each has limitations, and each requires practice to operate effectively. A group intending to use such equipment is well advised to practise with it before going on site. A final reminder: all three items can be handled with delicate care and precision—but carelessly handled can do irremediable damage to delicate objects.

SAND DIGGING

When using any of this equipment to dig down into soft sand, be prepared to move a great deal more than you may think is necessary. Sand is a singularly movable material and will slide back into holes with great enthusiasm, probably as quickly as you can suck it out. The stability angle of sand is rarely greater than 45 degrees and frequently much less. In other words, once the hole has been dug so deep that the sides are steeper than 45 degrees, you will almost certainly be fighting a losing battle. The area of the hole, at the sand surface, will inevitably become greater as it is deepened. Bear this in mind when deciding on a starting point, because it could cause instability of heavy items which may initially seem to be well away from the point of entry. In other words, ensure that heavy boulders, etc are moved well away from the area of excavation, so that they cannot suddenly roll down into the hole on to the operator.

We have now discussed the physical and mechanical methods of uncovering objects that we find on the seabed and should logically go on to methods of raising them. Before doing so, however, I will mention one other method of freeing objects, with which I freely admit that I have no experience—this is the use of explosives. Because of my own lack of experience and indeed that of the vast majority of divers, I do not intend to go too deeply into this method.

A very great deal of criticism has been voiced over the use of explosives on wrecks of archaeological importance and the present writer has certainly been as vocal as others. I deplore the use of explosives aimed at the quick recovery of artifacts, when it may result in untold damage to other items of, perhaps, less monetary value but of archaeological or historical worth. But I am willing to agree that in the hands of a skilled operator, plastic explosives can be so used that little, if any, damage is done to quite delicate items.

The secret is the complete knowledge of the power of explosives, knowledge of the lines of force of the explosion, and knowledge of where to place them. The correct quantity of explosive, judiciously placed along existing fracture or fault lines, eg the joint of the concretion to the iron barrel of a cannon or the joints between two blocks of stone, will achieve the separation of the items with no damage to either part. However, this is a skill that only a select few have (some of them might hesitate to display it in public!). The supply of explosives is strictly controlled and the users have to be registered with the police or have some form of licence—so relatively few will have access or ability. Those who have should use explosives only when absolutely essential and—please—with due care and attention to archaeological items.

RAISING ITEMS FROM THE SEABED

One thing I have noticed is the inability of most divers who are new to working under water to assess the weight of various items they pick up. So many times I have seen a man struggling back to the surface with a piece of amphora in his arms, probably quite unaware of the fact that it is full of mud and sand. Even when he eventually reaches the surface and has inflated his life jacket, he is still a liability to the team leader. Far more effort is required than is anticipated and the resultant exertion, especially at the end of a longish dive, can be a serious factor in diver safety.

Obviously, if we depend upon physically carrying items back to the surface we are not going to get very far. So we have to provide methods of raising our finds by other means. These can be self-contained or boat-mounted. Let us examine the latter first.

BOAT-MOUNTED LIFTING DEVICES

The simplest of these is just a line dropped over the side to which a bag, basket, or platform is tied. But very little weight can be safely lifted in this fashion, so it will be of value for only a limited amount of work. Nevertheless, if articles found when using an air-lift or dredge, are put into a basket as they are located, it will be simple to tie a line to this and haul away. Of this 'technique' little can or need be said. It is suggested that whatever is lifted is designed or modified so that any water contained within it will flow out as soon as it is clear of the surface. Even an old bucket with holes punched in the bottom will suffice, a strong net bag is ideal, but a kitbag, for instance, is not too good because it will contain a far greater weight of water than of artifacts, all of which must eventually be lifted over the side of the boat.

For heavier articles some form of boom should be contrived

Underwater excavation

and used. By far the most satisfactory is an 'A' frame consisting of two stout lengths of timber, the bases of which are possibly 1m apart, the tops being lashed or bolted together. At the apex of the 'A' frame is suspended a block over which a line is passed. With this type of frame, there is extra strength and greater stability. It is often easier to rig; for instance, it can frequently be mounted at a fairly low angle resting on the transom of the boat and projecting aft. A single boom would require shrouds to locate it and add support. These, to be effective, would take up far more room than the 'A' frame itself, also the actual boom would have to be twice as strong to carry the same weight.

Obviously, if the craft is large, then it may be possible to use a davit or a boom secured to the mast, both of which will facilitate swinging heavy objects aboard, which cannot be done with an 'A' frame. For really heavy articles a winch will be required, but surprisingly heavy articles can be manually lifted with multi-sheave blocks which can be obtained from chandlers or government surplus shops. These are used in pairs, one secured to the apex of the boom and the other to a strong fastening on the artifact. The line is fastened to the block at the boom head, goes round a sheave in the block on the object, back to the boom-head block, over another sheave and back down to and around the remaining sheave on the object block and returns up to and over the remaining sheave of the boom block and thence back down to the deck. Ensure that there is a bollard or something around which you can belay the free end of the line to help take the strain. Theoretically, by passing the lifting line over four sheaves as suggested above, the lifting power is increased in the ratio 4:1, so where we may have been able to lift 50kg on a single-sheave pulley, we can now lift 200kg with the same effort.

FASTENINGS (p 228)

Whatever the method of lifting we can use the same types of couplings to the objects to be lifted. These should be chosen with a view to the saving of time and effort under water. For instance, a rope could be tied around whatever is to be lifted —but to tie a good knot under water is rather more difficult

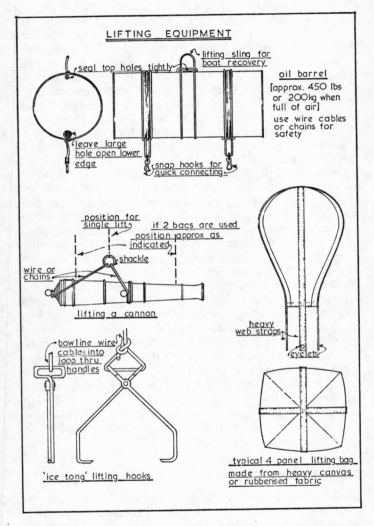

than doing the same thing on land, also the ropes will probably be of nylon or 'Terylene' and these are not the easiest to tie. I

Underwater excavation

would recommend whipping a bight into the ends of all lifting lines and fixing to these a snap hook and, if heavy loads are anticipated, using a swivel between hook and lifting line.

For light loads, the hook can be snapped on to the handle of a pail, or whatever is used to hold loose artifacts; it can be snapped back on to the lift line forming a slip noose around suitably shaped objects, for instance the necks of amphoras or the shaft of an anchor, or it can be engaged in slings or chains which are fastened around very heavy objects.

I strongly recommend the manufacture or purchase of lifting slings. In their simplest form they need only consist of a length of rope into which bights have been spliced at each end. By slipping one bight through the other a slip noose is made which can be placed around the object. Alternatively, the sling may be wrapped a couple of times around the object so that both bights are together and the snap hook engaged through both. A useful length for these is 1m overall and it pays to have several of this size available for work. For very heavy or irregular shaped objects such as heavy cannon, or masses of concretion, wire or chain slings should be used, the ends being joined with a shackle through which the snap hook may be engaged. Very great care must be taken when securing slings to a heavy object to ensure that it cannot slip while being lifted. Concretion can be brittle under load and if a piece were to snap off during lifting, it could allow the whole sling to move and the resulting imbalance might cause the load to slip from its lifting cradle. A safety precaution before actually lifting any heavy load, is to raise it free of the seabed for perhaps 1m and hold it there for a moment to settle before commencing the final lift, so allowing the diver time to inspect the fastenings.

Cannon were designed to balance, or nearly balance, on their trunnions, with a small bias of weight towards the breech. Therefore when lifting these, one sling should be located just in front (muzzle side) of the trunnions and another towards the breech. It may be possible to fasten around the pommellion, but it is not really advisable, for if much corrosion has taken place, the narrow neck of the pommellion could be greatly weakened. Again, carry out a couple of trial lifts for perhaps a metre to test for balance, for the load should stay level and must not be allowed to slip. If that should happen, even if there were no risk of it slipping completely out of the slings, it would be that much more difficult to bring aboard the lifting boat.

Obviously a code of signals is necessary between diver and boatman; in the case of a light line for small loads, a few tugs will convey the instructions. When using heavier lines, it will pay to have a separate signal line weighted as a shot line, on which signals may be passed. Fix the shipboard end to a bell, so that the boatman can be getting on with other jobs until the diver rings the bell to signal that he is either ready to lift or wishes to pass another signal to the boatman. It may sound silly, but another bell at the diver end, can be helpful at times. If weather is blowing up badly and heavy loads are being tackled, the last thing you will want is to be lifting in a heavy swell or for that matter to have your diver down below, so some means of attracting his attention is invaluable.

'Ice-tongs' can be used successfully, but I would advise against them if lifting in a swell. The security of their grip on or around an object is entirely dependent upon the weight they are carrying. It has been known for them to lose their grip on breaking surface in a strong swell, for although the load has been firmly held all the way up and even after clearing the water surface, a wave coming along will take some of the weight and that, plus the swell and surge could be enough to slip the tongs. In calm conditions they are fine and simple to use; no knots or special slings are needed, just engage the hooks around the object and lift away—but do be careful in rough seas.

In tidal waters, with heavy loads, it is possible to use the boat itself to lift. Make your lines fast to the object to be lifted and to the boat at low water, and, as the tide rises, so does the craft and the object—assuming that it is not heavier than the displacement of the craft! This principle is used in marine salvage, although there pontoons, not the salvage vessel itself, generally support the weight. As soon as the object has been raised clear of the bottom, the ship tows it into shallow water until it grounds again. On the next low water, the slack is taken up on board. The next tide raises both ship and load and the manoeuvre is repeated, and so on until the load is in very shallow water. The *Wasa* was towed to shallow waters in this way, before being loaded onto a dry dock.

Underwater excavation

SELF-CONTAINED LIFTING DEVICES (p 228)

By this is meant devices such as lifting bags or buoyancy containers, which are taken down either deflated or full of water. Once secured to the object to be raised, compressed air is fed into them, which either inflates the container or expels water from it and so provides buoyancy.

Dealing with the most unlikely type first, but always bearing in mind the possibility that you may be forced to use them at some time, we have metal containers such as oil barrels. Generally these have a capacity of 200 litres of oil (45 gallons) and in consequence have a lifting capacity of approximately 203kg (450lb). In many ways these are quite ideal for lifting very heavy objects for they are extremely strong and the raised ridges that are used for rolling the barrels along the ground, are very handy for locating lifting slings.

Their size alone means that problems of handling may arise and some practice is necessary before commencing any serious lifting operation. Getting them down to the seabed requires judgement and skill. The tare weight is likely to be approximately 25kg, therefore if the barrel is completely filled with water, the weight will cause it to plummet to the bottom. Fortunately, most barrels have at least two bung holes and sometimes three, the larger being approximately 70mm and the smaller only 25mm in diameter. The larger holes are located on one end and the side and the smaller is also located on the same end but usually at the opposite rim to the larger hole. See that the side bung and the smaller of the end bungs are very firmly screwed home, for these must not be allowed to come loose.

By rolling the barrel over in the water until the larger hole is level with the water surface, it is possible to effect a controlled filling of the barrel. Allow sufficient water to enter to provide a neutral buoyancy and the barrel can be safely guided down to the seabed. As each barrel has a lifting capacity of approximately a quarter ton, it is essential that the securing slings are properly constructed and secured to the barrel. This is best done before putting the barrel into the water. Make up the slings with two bights, one plain and the other with a snap hook, if available. In securing them to the barrel, do so with the snap hook bight on the same side as the large bung hole, so when an object is being lifted, the bung hole will be at the lower edge of the rim. Air can then be allowed into this hole from a hose and will rise to fill the barrel, expelling the water through the same hole. The air obviously will not be able to escape until all the water has been removed. (See p 234.) The rate of admission of air into the barrel can be carefully controlled and buoyancy accurately adjusted to whatever is required.

The limitations in using barrels are fairly easily recognised, for their potential lift capacity is so great that the object being lifted could get out of control, as we shall discuss a little later on. Also they are rather cumbersome and might cause damage to whatever is being raised. Finally, they can only really be handled from quite large craft. However, they may come in handy at some time, so it is as well to be familiar with their characteristics. Do be most careful in making fast the slings should you have to use a barrel, for if it came free from its load, it could cause untold damage to diver or boat as it rushed to the surface.

LIFTING BAGS

The conventional lifting bag is pear-shaped in design and usually made from either very strong canvas or a Neoprene fabric. The lifting lines are generally broad web bands that pass right over the top of the bag and are securely sewn to it for their entire length of contact. These end some 30–50cm below the open mouth of the bag which is at the narrow end of the inverted pear. Such bags have lifting capacity according to their size, are easily handled, and take very little space when not in use.

The bag is made pear-shaped so that air can escape, but it will come from a narrow opening directly under the centre of the bag and cannot therefore upset the balance or stability of the device. Additionally, the load is less likely to swing about, again because it is secured immediately under the centre of gravity. In principle there are no problems in using these bags; in practice, it pays to experiment before starting a serious lift-

Underwater excavation

ing operation, especially if more than one bag is being used on a particular object. It is always a little difficult to fill two bags at precisely the same rate and to the same degree of buoyancy, consequently if a cannon, for instance, were being lifted, the slings securing it might start to slip because of unequal lifting, with the cannon sloping down towards one bag. Practice with a crowbar, or other length of heavy metal, is a good idea.

Having taken the bag down to the object to be lifted, inflate it sufficiently to provide some buoyancy and make fast to the object by snap hook to the slings that will have already been put into position around it. When all is ready, gradually feed air into the open neck of the bag. Usually this can be done quite adequately, by taking out your mouthpiece and holding it up into the open neck. As this can easily be contrived to be higher than your demand valve, air will gush out and quickly inflate the bag. Try not to let too much in at one time, but fill in easy stages, checking to see whether there is any sign of movement. Frequently weight accounts for only part of the resistance to movement: suction from the surrounding mud and sand will tend to hold an object down, as will perhaps a slight adhesion from concretion. Once these are broken there may be too much buoyancy and the object will rise too rapidly for safe control. Always use only sufficient air to provide a very slight positive buoyancy at the bottom. As the bag rises, the air in it will expand as the water pressure drops and increased buoyancy will result. If the bag is allowed to rise too quickly it can almost jump out of the water on reaching the surface. The sudden loss of lift can cause the fastenings to slip and the object being lifted can easily fall back to the bottom. There is also risk to surface personnel and craft.

The problem is accentuated if the lift capacity of the bag is well over the weight of the object to be raised. Obviously one cannot get bags for every conceivable weight, but do try to use that which is nearest to the weight requirements. The reason is this. The total capacity of a bag will determine its maximum lift capacity, ie a bag full of air might displace water equivalent to a weight of, say, 100kg. If it is attached to an object weighing 100kg and filled with air, it will not lift off until the bag is full. As the bag and its load rise, so water pressure decreases and the air within the bag expands. The surplus air spills from

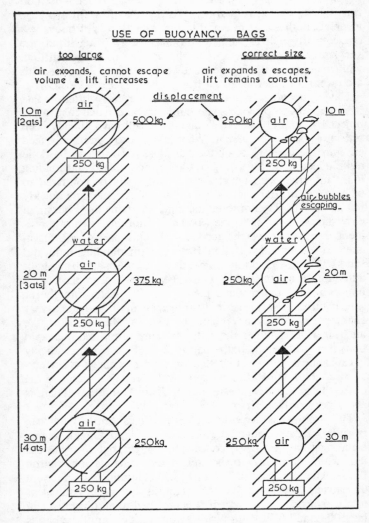

the open neck of the bag, still leaving a full bag capable of lifting 100kg and attached to 100kg load. So the buoyancy remains

Underwater excavation

positive and under perfect control, irrespective of the changing water pressure as the load rises.

On the other hand, suppose a bag which, when full, displaces water equivalent to 1,000kg, is attached to an object weighing only 250kg and at a depth of say, 35m. On commencing to fill the bag, the object would become buoyant when water equivalent to a load of 250kg had been displaced—but the bag would only be one-quarter full. The bag and load would start to rise and after some 10m, the ambient water pressure would have fallen by one atmosphere and the air in the bag would have expanded accordingly and would now be sufficient to lift nearly 350kg. Another 10m and the bag would still be only half full but would be capable of lifting 500kg and still with no chance of the surplus air venting or spilling out of the bottom opening. Yet another 10m and the air would have doubled again and the bag would be completely full and capable of lifting 1,000kg, but with a load of only the original 250kg. It would now be rising far too quickly to be controllable and any diver trying to stay with it would almost certainly run the risk of an embolism. Above it, any craft or swimmer that might be in the way would be very much in jeopardy (fig opposite).

This explanation should make it clear that lifting bags should be related to the load that they are to lift and that the smallest size for the job should always be used. It might be possible, with fairly small bags, for the accompanying diver forcibly to squeeze air from them as he rises, but this is out of the question with the larger sizes.

One can work out the lift capacity, as a guide, by calculating the volume in cubic feet and multiplying by 64 for salt water (60 for fresh); the answer will be the lift in pounds. A table of lifting capacities of spheres up to 450mm (18in) was given on page 216; the following will extend the range into the sizes needed in lifting operations. As our bags are not spheres, the capacities will not be identical to the lift capabilities of the bags; for a given diameter it can be assumed that a bag will lift approximately 20–25 per cent more than a sphere due to its being pear-shaped.

Salt Water Displacement of a Lifting Sphere

diameter		capacity	
mm	(feet)	kg	(cwt)
610	(2)	101	(2)
912	(3)	406	(8)
1,216	(4)	1,016	(20)
1,520	(5)	2,032	(40)

If purpose-made lifting bags are not available, kitbags can be quite successfully used for small objects. But bear in mind that a quite small kitbag measuring say, 250mm x 750mm will be capable of lifting 40kg when fully inflated. But would the securing lines and the fastenings to the bag be equal to this load? In all probability, the lines would be secured through the eyelets let into the rim of the bag, but it is very unlikely that they will carry a load of 40kg with complete safety. So have a care; it would be far better to use two bags, each secured to the same object and only partially inflated, rather than one fully inflated. It is possible to restrict the lift capacity of a kitbag by screwing the closed end into a tight roll and tying it firmly, so leaving only the desired amount of lift capacity.

Some objects are capable of acting as their own lifting devices, for instance, amphoras that are intact may be lifted by putting air into them while held inverted. This will not always be sufficient to make them completely buoyant, perhaps due to their being partially filled with sand, but it will aid lifting to a great degree. Whenever lifting in this fashion do ensure that there is an attendant boatman waiting for them, for if tipped open side up, they would immediately lose their buoyancy and would crash down to the bottom again.

In all lifting operations, it is necessary for the boatman to have lift lines which the surface snorkel swimmer can immediately attach to the slings fastened to the object. Snap hooks are invaluable here and even long poles with hooks firmly attached to the ends, can be successfully used from the boat itself. Without such precautions, some objects will almost certainly return to the bottom and probably disastrously! If one is called upon to lift a large number of small, fragile artifacts, for instance

Underwater excavation

pieces of glassware, jewellery, or soft timber, some means of preventing them from knocking against themselves and the container is required. I would suggest pieces of soft sheet plastics foam, probably about 25mm thick, cut into shapes suitable to fit, and laid flat, within the lifting container. If these are soaked in water before being placed in the container, they will require very little weight to prevent them floating away. On the bottom, the diver takes them out of the lift and places them on the bottom alongside, using the same weight to prevent them floating away. As finds are made, they are placed within the receptacle between successive layers of the foam plastic. This will prevent them bumping together, because the soft plastics surrounds each piece and the intermediary layers will separate on the vertical plane. As each find is placed into the receptacle, its marker is placed with it, or miniature markers if these can be provided, so that on arriving at the surface, they can be extracted in the reverse order of finding, but each clearly numbered to tie in with the site plan.

12

DATING AND IDENTIFICATION

EARLIER WE have spoken of the need for preliminary research to guide us to our site. Having carried this out carefully, we at last find ourselves working on a wreck and are finding objects. But are we sure we are on the correct site? Until we can positively identify it by indisputable facts, we cannot be really certain, for frequently there will be the remains of several wrecks in any given area. So what can we look for that can tell us something about the vessel? Obviously if we can even determine its age and size, we can at least decide whether it is a type similar to that which we are seeking.

There are many items on board a ship which provide clues to its size, dating, and possibly origin, but which, if any, of these are likely to have lasted sufficiently intact to provide the information we need? It would seem logical that the larger they are, the greater the chance of finding them, so our thoughts turn to the two largest items that may be expected to retain their form and be capable of providing information ... cannon and anchors.

CANNON (p 268)

Of the two, cannon can yield by far the most valuable information. Size alone will give a good indication of the size of the vessel. A ship armed with cannon powerful enough to hurl, say, a 32lb ball, would have been very large, probably a 1st, 2nd, or 3rd 'rater' (ie carrying 100, 90, or 80 cannon respec-

Dating and identification

tively) and would date from the eighteenth century. Also she would almost certainly have been a British ship, for few foreign ships, even Spanish ones, carried anything as large as this.

In describing armament, we will refer to pounds and inches rather than metric measurements, for all old records are set out in these units when describing weight of shot or size of bore.

In the descriptions of most old vessels, the number and size of cannon were invariably listed, and Admiralty records catalogued naval ships by their fire power, so it is obvious that the more that can be found out from the cannon, the better the chances of identifying the wreck. To help the reader judge the size and type of cannon, listed below are details of shot weight and dimensions of the cannon that would have fired it.

Poundage	diameter of bore	diameter of shot	length of barrel
48lb	7.34in (185.0mm)	6.99in (177.5mm)	8ft 6in (2.592m)
42lb	7.01in (178.0mm)	6.68in (168.8mm)	10ft 10in (3.050m)
32lb	6.41in (162.4mm)	6.10in (154.9mm)	9ft 6in (2.879m)
24lb	5.82in (147.0mm)	5.54in (137.9mm)	9ft 0in (2.745m)
18lb	5.29in (134.5mm)	5.04in (127.5mm)	9ft 0in (2.745m)
12lb	4.62in (116.6mm)	4.40in (102.6mm)	9ft 0in (2.745m)
9lb	4.20in (102.1mm)	4.00in (101.6mm)	7ft 0in (2.135m)
6lb	3.67in (93.7mm)	3.50in (88.9mm)	7ft 0in (2.135m)
4lb	3.20in (81.2mm)	3.05in (77.4mm)	6ft 0in (1.829m)
3lb	2.91in (73.7mm)	2.77in (70.8mm)	4ft 6in (1.371m)

Of the dimensions given above, those relating to bore or shot size are the most indicative of the gun size, for length did sometimes vary and later cannon were frequently shorter for the equivalent poundage. The different sizes of cannon carried different names such as Culverine, Saker, Falcon, etc, which can be found in old records and in books listed in the bibliography.

In assessing the number of cannon associated with a particular site it is well to remember that those found may only represent a proportion of the original armament. Some may have been either deliberately or accidentally jettisoned some distance from the wreck site or may even have been already salvaged by earlier divers. However, tabulate the number and sizes of those found and compare this figure with the Admiralty (if naval vessel) or owner's description to see whether this tallies with the vessel being sought.

Fortunately for us, even one cannon can frequently tell us all we wish to know. Most cannon were date stamped when cast, primarily for the benefit of the armourer whose duty it was to ensure that all cannon were serviceable. Knowing the date of original casting, and the amount of shot fired through it, by examination, he could reasonably judge the condition of the bore.

The date would normally appear either in full on the reinforcement nearest the breech or, more frequently, as a four- or two-figure number on the ends of the trunnions. If only two figures appear, these refer to the year and not the century, ie 1787 might well appear as 87 only. This can lead to uncertainty as to which century should be applied and additional identification must be sought. Frequently the opposite trunnion would show the maker's mark or initials and this can help; also in the case of British cannon the reigning monarch's cypher would often appear on the first reinforcement and this would identify the century. The Dutch East Indiaman *Amsterdam* had the letters VOC (Verenigde Oestindisch Compagnie) formed into a monogram on the first reinforcement and this, together with the date of casting, identified her without any doubt.

The formation of the numerals also provides a clue. Most British cannon of the seventeenth and eighteenth centuries had the numerals embossed or raised, both on the reinforcements

Dating and identification 241

and trunnion ends and only in the late eighteenth century did an additional series of numbers come into being. These were four-figure (usually) numbers and were cut or incused into the casting. Occasionally a four-figure date will be found which is also incused, but common sense will generally decide which is the correct interpretation. I wish it were possible to be really specific so that one could arbitrarily say that such and such a type of marking meant a specific date span, but unfortunately there are too many exceptions to the rule to allow this.

The markings on cannon, particularly the embossed type, can be severely disfigured by sea action and this drives home the need for care in removing any extraneous coverings there may be, in case careless action should still further damage important markings. More will be said about this in the next chapter.

ANCHORS

These can range from the very early stone type to modern stockless and CQR models. Unfortunately, there was relatively little distinctive change throughout the range of iron anchors from about the fifteenth century to the late nineteenth or early twentieth centuries, so they do not provide much assistance in actually dating a wreck. However, their size provides a good clear indication of the size of vessel to which they had been attached. Early records listed both the number and weight, sometimes also the size of anchors for ships of various tonnages. It is interesting to note that frequently the diameter of the mast was used as a 'yardstick' when referring to the appropriate anchor. Logically the diameter of the mast varied with the size of the vessel and those of similar displacement probably had masts of very similar size. These scales were laid down by Admiralty for warships and can be reasonably taken as a guide for merchant ships of similar size.

Another interesting feature is the sheer number of anchors carried by those ships, frequently as many as twenty-four on one vessel. But when it is realised that if the fluke of an anchor became wedged in a crack in a rock, it was quite beyond the power of the crew with their straight-pull capstans to free it, the number is explained. In times of emergency, anchor lines were often cut, for even under ideal conditions it could take a crew a couple of hours to winch in one anchor. If a storm were approaching, or an enemy for that matter, it could be necessary to sacrifice the anchor, so losses were inevitable, hence the need for replacements.

The only real sign of change in the design of iron anchors was the replacement of the wooden stocks (the cross-bar at the top of the anchor) by iron ones and this appears to have taken place during the late nineteenth century. More detailed changes occurred in the cross-section of the shank, both in shape and formation, and also in the shape of the flukes, but these do not appear to have been chronological changes so much as local design changes, so do not help in general dating, though they may be of assistance in determining the place of origin of the anchor. Reference to publications listed in the bibliography may well be of further assistance in this direction.

TIMBER

This is of relatively little value for dating or identification—certainly most British ships were made of oak, but some were of teak and by the same token many Spanish ships were also built of oak, so there is nothing conclusive here. Similarly pine-built vessels may have come from either Scandinavia or the North Americas, so this is no real guide either. In any event, much of the wood will have vanished and what little is left will require some knowledge of timber to identify it, so I feel that on the whole little information can be gained from any wood which may be found.

FASTENINGS

Metal nails have been used for centuries and apart from metallurgical examination to determine the alloys used (mainly applicable to vessels of comparatively recent construction), little date evidence can be obtained. I have found bronze nails lying in the sand at Bithia, Sardinia, possibly dating back to the fifth or sixth century BC, which are very little different in shape to some I have found still embedded in a baulk of timber from a wreck of fifty years ago.

Dating and identification

Iron nails were used predominantly during the middle ages and it was not until the latter part of the eighteenth century that any radical change occurred in the design of nails. Before that time most nails were wrought or forged, but the Industrial Revolution brought machinery with which rolling techniques could be employed and nails from that period onwards generally had round sections. About this time copper nails came into use above the waterline, replacing the earlier iron fastenings, so this provides another guide. Still later, probably about the late nineteenth century, changes occurred in the metal used; bronze alloys came back into favour and points also became flattened or 'chisel' ended. This type is still in use today.

Underwater parts of the hull have generally relied upon trunnels or tree nails for securing planking and ribs. Wooden pegs or fixings were driven into pre-drilled or augered holes in the timbers and the water caused them to swell, thus tightening up the whole structure. This practice has been in use from the earliest times and the only variation appears to have been in the method of producing the trunnels and any attempts to date by this are likely to be very inconclusive. The very earliest were merely split lengths of wood which may have been roughly rounded by knife, but on which the original split lines would frequently show. Later the use of spokeshaves or drawknives produced a better roundness, with narrow-cut facets all round. The late eighteenth century saw the introduction of mechanical methods of producing trunnels, or dowels as they are now known, by forcing wood through holes in metal plates. This left scratch marks running the length of the trunnels as evidence. Still later, in the nineteenth century, turned dowels were used in some cases and the marks of turning can be seen on these. But they were comparatively slow to produce and were not in widespread use.

Another significant date line is the introduction of copper sheet for sheathing the underwater sections of hulls to counter-act shipworm or *teredo* and this came into prominence during the latter half of the eighteenth century. The *Amsterdam*, for instance (1748) was not clad with copper, although narrow strips of copper have been found backing the joints of the planks where they joined the stem. This measured approximately 100mm wide and ran vertically down the stem. So this was evidence of the coming use of copper sheeting. Any hull which is sheathed in copper must date from the very late eighteenth century and if a merchantman of not too great an importance, from after the turn of the nineteenth century.

It is not uncommon to find marks of small nails, if not the nails themselves, which might indicate the use of copper sheet, but be careful: they may only be indicative of the fastening of a patch, possibly of lead, not full-scale sheathing. Occasionally padding would be nailed to areas of planking to prevent chafe damage to ropes etc, and the nail marks would still be left. Additionally, lead and pewter were used to face scupper holes and would have been secured with nails. Finally, shipbuilders of old realised that iron oxidised and formed into the hard iron-pad surface and they used this fact to advantage by close-nailing vulnerable areas, such as the sternpost, keel, and stem, with short but broadheaded nails. The iron pad quickly formed a close protective covering over the areas concerned and was a major deterrent to shipworm. In other areas, also vulnerable to attack, the constructional timbers were covered with a mat of cow hair and pitch which formed an impervious coating through which the shipworm would or could not penetrate. This was held in place by thin planking which was attacked but could easily be replaced. It can thus be seen that nails or nailholes did not necessarily indicate copper sheet, so be careful: do not jump to conclusions.

COINS

Here we have a most valuable date indicator. It is very uncommon not to find some coins on a wreck site, for, small though coins are, their weight carried them into crevices and kept them in the site area. Relatively unaffected by corrosion, they can frequently be examined and dated. One coin may not be indicative from a dating point of view, but if a number are found of a similar date span, then it is reasonable to suppose that the wreck occurred shortly after the date of the latest. For, just as today, the mint produced many coins of each denomination each year. However, just one coin can be misleading for it may have been carried by a crew member for

Dating and identification

luck or it may have been in circulation for an excessively long time. By the same token, the country of origin of the coin is not significant if only one is found; it may have been a souvenir.

It should also be remembered that many a warship carried prizes taken from a ship of another nationality, so even an abundance of one particular denomination is not necessarily conclusive. To make matters worse, many coins of gold and silver were used almost as international currency, so although they can help considerably in dating a wreck, their presence on a site should be interpreted in conjunction with other artifacts before positive conclusions are drawn.

WEIGHTS

Under this term is understood those weights which were used on scales for weighing bullion and foodstuffs, etc. Strict control of the accuracy of scales was exercised in most countries and all weights were periodically inspected, tested for accuracy, and date stamped. Although this was not done at regular intervals, usually only when a ship returned to its home port, the frequency was such to ensure a date stamp every 2–3 years. As weights were made of brass, they were not unduly damaged by water and when found can frequently give both date and nationality of testing office.

POTTERY

One of the blessings to an archaeologist is the way in which fired clay is preserved. Throughout the world innumerable examples of pots, containers, bowls, jugs, etc have been found, which have been very adequately recorded, catalogued, and dated. Even prolonged submersion in sea water has little effect, as is evidenced by the number of amphoras pulled up in the Mediterranean where they have been for hundreds if not thousands of years. The amphora can date a find, identify its point of origin and, through its contents, tell us much about the type of trade in which the vessel had been involved. The very wide variety of types have been catalogued thoroughly (see Dressel), and quick identification can be achieved by checking finds against these charts.

Well after the Romans and their amphoras, earthenware fortunately continues to be a source of information. The Spanish used large containers which are generally known as 'olive jars' and these were as ubiquitous to them as the amphora to the Roman and the jerrycan to the twentieth-century traveller. Equally specific records have been compiled of these containers, indicating both type and date span, so again one find referred to such catalogues will provide a great deal of information. It is not essential to obtain a complete, undamaged jar to secure identification; experts in pottery can tell from colour, type of firing, and method of production of even small pieces, from what era and source it came. Details of surface design and patterns will help date to probably within fifty years—so do not throw away even the smallest piece without first showing it to an expert.

TOBACCO PIPES

During the early part of the seventeenth century, smoking of tobacco became widely accepted and seamen were no exception. In fact, they were probably more inclined than any to the habit, for it was believed to be a protection against disease as well as being a solace. Most ships carried large quantities of pipes packed in boxes or bags surrounded with buck wheat to protect them from damage. Clay pipes were very delicate affairs and even a sudden turn of the head would be sufficient to cause the bowl to part company from the stem! So it is not surprising that large quantities were carried at sea. As in many other items of a fairly frivolous nature, design varied considerably, according to regional influences and time of production. These various designs have been exceedingly well recorded and once again comparison of a pipe bowl found on site with published catalogues will quickly provide a clue to place of origin and date of manufacture. Additionally, it is interesting to note that many clay pipes had small projecting pieces at the base of the bowl, presumably to keep the hot bowl off table tops, and on the ends of these projections can frequently be found the maker's stamp or initials, yet a further source of information.

BOTTLES

Here again we have a form of artifact which withstands submergence in the sea very well and if found in a reasonably whole condition will offer clues as to both date and origin. The gradual change over the centuries of bottle shape, size, and glass composition is well recorded and experts can date a specimen to within a matter of 30–40 years. Generally there is a fairly clear-cut demarcation between the different periods of bottle design for, since they were of a comparatively fragile nature, their expectation of life was not all that long and in consequence it is a rare thing to find any quantity of bottles together on one site which date from radically different periods. Although one has to be on guard in case a consignment in a particular vessel contained 'vintage' wine in addition to the contemporary brew!

Frequently bottles, certainly of the former, would bear the maker's or shipper's seal, complete with date and name—so watch for these. It is also possible to examine the contents and deduce their place of origin. Bottles found on the *Amsterdam* have been established as originally having been filled in France and are likely to have been a Bordeaux wine—so perhaps the contents may only serve to confuse the issue!

There are many other items that may be found on a wreck which can also help to identify or date it, ranging from personal possessions (on one site a watch case was found in which the maker's name and date were inscribed), through objects found in the cargo to general equipment. So the rule must be to note every find with great care and to spurn nothing in your search for clues. However, those mentioned above are the more easily identified and I trust will be of value to you. In the bibliography I have listed various publications which go in greater depth into many of the items mentioned; your attention is drawn to them.

13
CONSERVATION AND REPORTING

IN THIS final chapter I would like to discuss what is probably the weakest aspect of underwater archaeology—the conservation of any finds that have been made and the adequate reporting, if not publication, of the work done. Unless we are capable of coping with the consequences of transferring our finds from the watery environment in which they have been steeped for so long, to the entirely different conditions of atmosphere, and unless we are prepared to make the very considerable efforts involved in writing up our work for others to read and assimilate, it would be better had we not commenced.

For until and unless both of these functions are completed, we have not been carrying out an archaeological project, we have merely been catering for our own interest and possibly gain. An essential part of archaeology as a discipline is the detailed recording of the finds as a whole within the context of the site and then the complete preservation of those finds. For it is only thus that a full interpretation can be made and conclusions drawn.

Treating conservation first, if only in the logical sequence of work, it can be immediately said that there is much to be learnt and precious few sources from which to learn! For many years there have been excellent books written covering conservation, but unfortunately mainly from the land dig aspect and it is only in recent years that the increase in underwater work has raised the problems of treatment of articles that have been exposed to water, especially salt water, and marine life.

Conservation and reporting

Work is currently being carried out in this new branch of conservation, but little has as yet been published.

Various archaeological institutions are striving to perfect new techniques and many local societies are also taking an interest in the new development, so it is quite likely that there is someone, not too far away, who has the facilities and some idea of how to set about the work. It behoves the expedition leader to find out where and who these people are and make arrangements with them for treatment of any finds. The golden rule then is to get the artifacts to them as quickly as possible, and leave their treatment to the experts! The suggestions on treatment that follow are mainly for interest and to indicate the problems that are involved, but, I repeat, the actual treatment should be left to experts wherever possible.

An object that has been buried in the seabed for a couple of centuries will not alter very much if left undisturbed for a few more years. But bring it up into the atmosphere and in a matter of days, decomposition will commence. Regrettably, already there are far too many iron cannon, cannon balls, anchors, etc, lying ashore which are rapidly deteriorating into a spongy mess. So, having said that the important thing is to get the articles to people with proper facilities, it is obvious that some 'first-aid' steps must be taken to preserve the objects if only to tide them over the time taken to get them to the laboratory.

The first beneficial treatment is to immerse finds in fresh water, so watertight containers are a must. Buckets, bowls, tubs, all can be used; even a hole dug in the sand or ground and lined with a sheet of polythene with the corners folded will make a suitable receptacle. A fresh-water stream is invaluable, providing the objects are securely retained in wire, plastic, or woven rush baskets to eliminate the risk of loss. The constantly changing water is excellent treatment. Should static containers be used, endeavour to change the water as it becomes brackish through the salt that has been leached out of the artifacts. Providing this action is taken immediately the items have been brought out of the salt water, little further harm will come to them. The treatment is the same whether they are animal, vegetable, or mineral!

For actual conveyance, the size and shape of the articles will dictate the nature of the containers, but they should be robust and if possible watertight for, with the possible exception of stoneware, the articles must be kept at least damp in transit. Glass bottles, such as the very large sweet or candy jars with screw tops, can be used but may need additional protection against knocks. By the same token, small objects placed within them risk damage in transit through knocking against each other. It is wise to fill these containers with damp sawdust with the items spread out in different layers, or single larger items may be well wrapped in wet paper (pulped newspaper is good) or wet cloth. Wood, leather, and less fragile items can well be contained in such bottles. It is also possible to buy self-sealing plastics containers ('Tupperware' for example) in various sizes and shapes. These are lightweight, not liable to breakage, can be quite water- or damp-tight, and are available from many shops. Food can be taken on the expedition in such containers which can then be put to use after the food has gone the way of all things!

Wooden or metal ammunition cases are excellent, but are best lined with sheet polythene to retain moisture. Alternatively, the artifact may be sealed into a polythene bag together with some water, and the whole lot surrounded by shock-absorbing materials. Crumbled cork, polystyrene, and sawdust are just a few of the packing materials that will prove satisfactory. Where such materials are in actual contact with an artifact they should be treated with a preservative such as formaldehyde (the aqueous solution is known as formalin). Formalin should generally be used at a maximum strength of 10 per cent. Various forms of benzene compounds may be used, and at varying strengths, so get advice on the correct rate from your supplier. As a guide, sodium benzoate is best used at only 2 per cent. Ortho phenylphenol sodium salt (1 per cent), thymol, and toluene can also be used. The purpose of these chemicals, added either to the packing materials or to the final rinse before packing, is to act as preservatives and corrosion inhibitors, also as fungicides to destroy harmful organisms.

Heavy items such as cannon should also be damp-packed in specially made heavy wooden cases. If too large for this, then completely wrap them in layers of sacking over which buckets of water should be regularly tipped to ensure constant damp-

Conservation and reporting

ness. If it is anticipated that this will be the only way of transporting them and that salt water may be the only water available to maintain wetness then do not wash them in fresh water at all, but leave them salty until such time as fresh water is constantly available. Much damage can be caused by alternating the types of water in which an object is submerged.

Frequently large objects, such as cannon, will be brought up heavily encrusted with calcareous deposits. It is wiser to leave this covering intact, even though the natural desire is to remove some, if not all, either to try and locate a date or identifying mark, or just to 'see what it is like'. If left, it is an additional protection to the metal. If removed damage may be done to the undersurface or protuberances that may be concealed. Frequently, the layer of metal immediately under such encrustation has been affected chemically, ie turned into oxides or sulphides; if the outer covering is removed prematurely then, on exposure to air, further chemical changes can occur which will inflict even more damage upon the article. So please do not remove any covering matter until full conservation facilities are at hand.

In the case of smaller objects, not infrequently the original content will have changed completely to sulphides, generally black in colour, or oxides. This incidentally can be a useful clue —if prodding into a mass of apparently natural encrustation produces a cloud of black in the water, this may well be iron or silver sulphide, indicating the presence of those metals. If a lump of encrustration appears to have a 'formal' shape it may well have been formed around an object and possibly the object will have been converted chemically to just a sludge, leaving a hollow centre. Careful cutting with a hacksaw or diamond wheel, into two halves, will produce the makings of a good mould for the object. Cut, additionally, a sprue or fill hole in one edge. Next cut a template of cardboard or paper just equal to the thickness of the original cutting tool, place this between the two halves, and tie them together. You will now have a female mould of the object. Fill this with plaster, or better still one of the special rubberised moulding compounds which will allow moulding around any undercut feature in a complicated mould, and when this is set, it will form a perfect replica of the original object. It pays to treat the inner surface of the mould with a varnish or release agent to prevent possible sticking of the cast to the mould (p 253).

When attempting to remove calcareous matter, do not, please, use a large mason's hammer, cold chisel, and brute strength! Untold damage can be done to the underlying surface and it is quite an unnecessary waste of time and energy— just gentle taps straight on to, and not obliquely to, the surface will result in the coating cracking along natural fault lines, such as the junction to the metal surface, and lumps will come away quite cleanly. Be always conscious of the likely presence of a protuberance such as a small handle, lever, catch, or even embossed numbers, which an over-zealous blow could well destroy. Generally such treatment will be reserved for substantial objects, but at times large amounts of small pieces of encrusted material will be brought up, all of which might well contain items of interest. Place the lumps on a smooth hard surface (an anvil is ideal) and tap directly on to them until they crack open, but again do be circumspect in how you do this—a heavy hammer on a delicate brooch does a lot of no good!

Generally on smaller articles or where the covering is very thin, a chemical treatment can be used. Apply nitric acid (10 per cent) by swab or brush, with frequent rinsing and inspection. In this way the covering can be gradually removed without marking the surface beneath. Another useful chemical for this work, especially on non-ferrous metals of the copper family, and rather more pleasant to use is sodium hexametaphosphate which can be obtained under the brand name of 'Calgon', the water softener. The pure chemical should be used at a strength of 10 per cent; 'Calgon' may be rather stronger in solution to get the required results. Hydrochloric acid (spirits of salts) is another chemical which will remove tarnish and some foreign matter, but also must be used with great care and frequent rinsing.

Patina and verdigris may be removed from very delicate articles, eg coins, by using a vibratory tool such as used for engraving names on dog tags. The tool is applied very gently, a large magnifying glass being used to observe progress. Very small areas are covered at a time, with constant inspection. But it is really wiser to leave such treatment to experts

Conservation and reporting

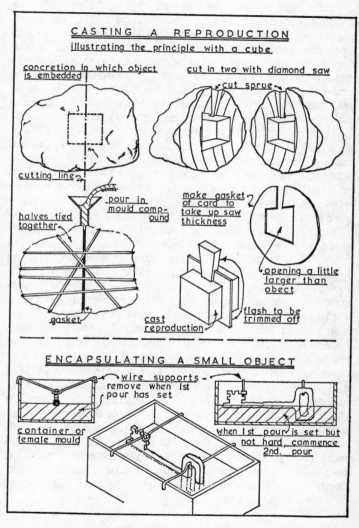

who have full facilities and can also help to identify and date the object.

In the case of iron and cast-iron objects, the possible presence of iron sulphide underneath a covering layer of calcareous matter and the advisability of leaving this covering intact, has been mentioned above. The situation is that although the surface of the metal may appear sound and in reasonable condition, it may not in fact be so for the iron sulphide may still be active. So long as it remains moist no change will apparently take place, but upon drying, the surface will crack badly and may crumble away. Time and time again one sees iron cannon balls with deep cracks penetrating right down through the surface. This is frequently due to imperfections in the original casting. At the time of their production, perfection in casting was not sought, for their essential purpose made this quite unnecessary. In consequence, line faults were common, corrosion would occur in such faults and, on drying out, would show even greater penetration in the form of deep cracks.

Frequently the exterior of iron articles is covered with another form of encrustation which appears as a rust-coloured matter in which all sorts of extraneous material, such as pebbles and shells, has become embedded. Yet another type to look for is the almost granular type of black crystal-like substance (known as ironpad) which is more generally associated with cast iron and consists in the main of graphite particles, which are in themselves almost indestructible and which have been bonded together by the iron sulphide. On the *Amsterdam* we found one very large patch down on the starboard bow covering nearly a metre square. This was tremendously hard and withstood some experimental hard blows without cracking although it could be flaked off by levering it free from the ship's side. Generally the removal of both of these forms of encrustation requires mechanical means, but soaking in fresh water will help considerably and judicious use of nitric acid will also produce results. By all means experiment yourself on simple and valueless objects, but in a case where the object might appear to have significant importance, you will be wiser to leave it to the expert.

ELECTROCHEMICAL TREATMENT

The following treatment should again be left to experts; if it is incorrectly applied irreparable damage may be caused. For those who wish to become fully conversant with the techniques, works are listed in the bibliography which discuss the subject more fully than is possible here.

Ferrous metal will respond to a treatment of an electrochemical nature which involves prolonged soaking in baths of fresh water to which has been added enough caustic soda (sodium hydroxide) to make a 10 per cent solution. Depending upon the size of the article the initial soaking could take anything up to two months, although heating of the bath and contents will reduce this. After approximately 4–5 weeks, zinc, in sheet or granular form, is placed in contact with the object being treated. In due course, the oxygen from the corrosion will be drawn out and will form a white deposit on the article in the form of zinc oxide. This is allowed to accumulate until no further increase can be noted, and then the article is washed down with a weak solution of sulphuric acid followed by soaking in distilled water. Where distilled water is not available, it is possible to use tap water (providing it has not been chlorinated) or deionised water. Any traces of chlorine could react with any ferric chloride that may have been formed and will aggravate the situation. If there is any likelihood of any chlorides being present, treatment is continued until none are left. This is checked by adding a drop of silver nitrate and nitric acid (a solution of 5–10 per cent for both). If a white cloudy discolouration occurs in the water, it is an indication of the presence of some chlorides and soaking must continue until tests show complete freedom from contamination. When this stage is reached, the article is dried as quickly as possible, using artificial heat if necessary. After a final sealing with paraffin wax, clear polyurethane varnish or even complete plastic encapsulation, there will be no possibility of further chemical degradation. During the washing spells, the water is changed frequently and plenty of it is used in each bath, as this speeds up the cleansing action.

ELECTROLYTIC TREATMENT

As an alternative to electrochemical treatment as already described, one can resort to electrolytic treatment. We have mentioned how articles found in salt water may have set up an electrolytic reaction with objects of a dissimilar metal that are either touching or are in close proximity. It is quite possible to reverse the process and by immersing the article in an electrolytic solution or bath, providing the core is sound, the salts that have been formed on its surface may be decomposed, leaving a clean article. This is easier with some metals than with others and is in any event a ticklish job which is better left to the expert. Faraday's laws of electrolysis (1834) set out the two main factors governing this process and point out that the amount of electric current passed through the solution and the chemical equivalent weight of the metal will control the activity of the reaction. Thus it is obvious that although the setting up of an electrolysis bath is relatively simple, a sound knowledge of chemistry and metallurgy is essential when one is handling possibly very delicate objects.

TREATMENT OF PEWTER

Pewter presents its own particular problems which, I believe, justify a separate paragraph. It is an alloy of lead and tin, the former of which will not be affected by exposure to salt water, but the tin can become severely attacked by the salt and is affected by cold. Frequently the surface of an object made of pewter will be disfigured by large bubble-like protrusions, generally known as 'pewter-pest'. The tin content may be chemically attacked to form a salt which appears as a silvery powder, or it may undergo physical change into a grey powder when reduced below 18°C for a considerable period of time. It is the breaking down and expansion into powder which forces up the surface into the bubbles.

Where this decomposition has occurred, very little can be done to remedy the situation. If the find is valuable and delicate, possibly the only way of arresting any further disfigurement is by encapsulating the whole object in plastics. If only a few bubbles are visible, try to open them very delicately with a sharp, thin file or sawblade and scrape out the loose powder from within. Follow this with a soaking in a fairly strong alkaline solution (potassium hydroxide lye for instance). This same soaking will help to remove any minor corrosion. After perhaps 20–30 minutes soak rub down lightly with very fine wire wool, lubricated with soap. A 'Brillo' pad is very good for this and after a short time, a black coating or film will appear, similar to that obtained with silver polish on silver or metal polish on brass. Wash this off and examine to see the progress and repeat as necessary. When satisfied with the appearance, rinse in warm fresh water, and dry. Meantime, be very careful to keep the mixture from your clothes, skin, and eyes.

CERAMICS

Of all materials likely to be found, metallic or otherwise, ceramics are the easiest to treat, especially the well-glazed articles—they need no treatment, at least for preventing unwanted chemical action. However, it may be necessary to remove calcareous matter and this is best done with nitric or sulphuric acid, but only if the glaze is sound. The latter acid will also remove rust stains which may occur through being in contact with a ferrous metal article. In general it is a good rule to soak any form of ceramic material, glazed or otherwise, in fresh water to remove the salt that the porous material will have soaked up. If glaze is flaking, a coating of polyurethane varnish will often hold the flakes in place if applied fairly fully with a very soft brush, virtually running the varnish on. No special form of drying is required after soaking; probably gentle sunlight or indirect heat is the best.

ORGANIC OBJECTS

When treating animal or vegetable materials, one major fact stands out and that is that they are generally more porous and will have soaked salt water right into their very fabric. The first stage of treatment is to ensure that every vestige of salt is leached out. This is achieved by prolonged (2–3 weeks) immersion in changing fresh water and this stage must be completed before commencing on the following steps. The article must then be literally dehydrated and this requires the use of hydrophilic substances, such as alcohol, and prolonged soaking in these will drive out the water. Once again the treatment should be progressive, with increasing strengths in each successive stage. Generally this can be achieved by having three or four successive baths, each of a greater strength. The length of time for which an article must be immersed in a particular bath will, to a great extent, depend upon the volume and porosity of the object. But with a range of baths, the objects may be moved on from strength to strength according to their needs.

LEATHER

The treatment described above applies well to leather and should be followed by soaking in a wax solution. To make this, take a suitable solvent that will dissolve paraffin wax. Gradually add more and more wax until wax crystals begin to form, indicating saturation of the solution. Allow the articles to soak in this for a length of time relative to their volume; a thin strap might, for instance, only require a couple of weeks, whereas a heavy shoe might need all of 6–8 weeks. Eventually take it out of the solution, allow the solvent to evaporate, which will leave a film of wax on the surface, polish this off with a soft duster or brush and a good surface will be left.

WOOD

We can follow the same treatment as above on small objects, but larger structures that may in their own right be weight supporting—I am thinking especially of the *Wasa*—require something more substantial and lasting than alcohol. Among the more successful ingredients is glycol, which is probably more familiar as anti-freeze for motor engines. As commercially sold for that purpose, the glycol is usually either mono- or di-ethylene glycol and contains additives designed to protect the engine from corrosion, also colouring to indicate a particular brand or just that the coolant in the system has been treated. For preservation purposes we need a chemically pure di- or polyethylene glycol without additives which could very adversely affect the article under treatment.

Wood has to be under water for a very long time before penetration is great. The *Wasa* and the *Amsterdam* have both shown that even after two or three hundred years below the seas, penetration is only in the region of 10 centimetres. Beyond this the timbers were their original light colour and, on the *Amsterdam*, smelt almost as though fresh cut! But this first few centimetres is of importance, especially with a small object or a surface on which carvings appear, where the loss of the skin surface could lose all. If allowed to dry untreated, the soft outer layer will certainly flake off; additionally, it will be very susceptible to attacks from fungi from airborne spores.

The glycol is introduced to ensure complete penetration of all the wood cells and to drive out the last traces of water. This ensures that shrinkage is minimised and that the weak cell structures are reinforced, an important preventative against splitting. Once again, progressive treatment takes place during which the strength of the glycol is steadily increased until sufficient has been absorbed into the cells to congeal into a firm mass. As glycol has a very wide temperature range, (it was used as a coolant in aircraft engines as well as anti-freeze), no possible temperature change due to climatic conditions can have any effect on the congealed glycol or the cells it occupies.

Small articles should be treated in a bath containing, initially, a 30 per cent solution of polyethylene glycol. This strength can be gradually increased by heating the mixture and allowing the water content to evaporate, topping up with similar proportioned mixtures until a strength of some 60 per cent is reached. A hydrometer will indicate when this stage is reached; mix up separate batches (quite small) of 30, 40, 50 and 60 per cent mixtures and mark, on the side of the hydrometer, the float level for each strength. Check the soak baths by regularly taking readings on the hydrometer and noting the float levels. It is also possible to use a series of baths, as before, each with an increased strength. This is often the better way for it is then possible to leave objects of varying sizes in a particular strength of solution for different times and so ensure full penetration. It is better to leave them too long than not long enough, so err on the generous side rather than the other. When treated as much as appears to be necessary, remove and heat gently, with a radiant heat source if possible, and wipe off the excess glycol which will exude from the article. Finally, wash down with a spirit solution to remove the excess which will otherwise become very sticky and dirty. Again gently dry and the treatment is complete.

GLASS

Glass is also, perhaps surprisingly, subject to reaction after submersion for long periods. The change is evidenced in the surface colouring which becomes semi-opaque and iridescent. Initially, this looks very beautiful, but if left untreated will become progressively worse, until eventually, the glass will commence to flake off in quite disastrously large flakes. Great care must therefore be taken when trying to remove any dirt or calcareous matter that may be on the surface. If the bottle is reasonably clean, it may be wiser to leave well alone, using a plastic spray or very soft brush and varnish to prevent the flaking from becoming worse.

If the bottle is still corked, take great care to examine it closely for it may bear the seal, or at least remnants of it, which might tell much about its origin and its contents. If the contents appear intact, it is possible to analyse them and gain some clue as to their origin—but don't be too optimistic about drinking; it is likely to be a very unpalatable drink. Bottles brought up from the *Amsterdam* were in many cases intact, but the contents were definitely undrinkable and on analysis proved likely to have been a French Bordeaux wine of indeterminate colouring. It appears that even the reddest of wines becomes a yellowy straw colour after a time. However, the alcohol content could be established and proved to be something in the region of 18 per cent. Salt water had penetrated the cork and it was decidedly a 'corked' bottle! Gin brought up in stoneware bottles was almost drinkable and clearly gin, and many of the bottles still bore seals. The type of drink in a bottle makes a difference; a gassy drink such as champagne is likely to withstand spoiling more than a non-gaseous drink. A comparatively recent wreck, about 25–30 years old, was found to contain a cargo of pernod and champagne, much to the delight of the divers. The champagne was found to be in excellent condition, but not the pernod. The difference appears to be due to the pressure built up inside the champagne (gassy) bottle which, in conjunction with the shape of the cork, withstood the pressure of water and kept it out.

ENCAPSULATION (p 253)

Encapsulation has already been mentioned as a means of preserving small and perhaps delicate articles, which it may not be possible to handle or treat in the more general way. It is a process of completely surrounding an object in a protective medium, in this case a clear, hard plastics material. Any of the acrylics will do, but perhaps 'Perspex' is the best known. An inquiry to any of the major manufacturers of thermoplastics, such as ICI, Monsanto, Shell, etc, will produce advice on suitable plastics, speed of hardeners, and probable local sources of supply. The clear plastic will come in liquid form with a fairly viscous consistency; its hardener or catalyst is also liquid but usually free flowing. Ensure that both containers are kept tightly sealed when not in use to prevent thickening due to evaporation. When the two liquids are mixed, a chemical reaction takes place which causes the mixture to go solid but at a definite speed according to the hardener chosen.

Choose or make a mould of a size convenient to the object to be treated, for although the plastics may be cut and shaped quite easily after curing and without injury to the contents, an unnecessarily large mould is wasteful of material. Glass or metal are ideal materials for moulds, but do not use plastics containers as these will almost certainly react with the hardener. In the case of large objects, it will suffice to knock up a wooden box for a mould. Hard plaster of paris will make good moulds which are best prepared by taking a female casting from a male mould or former.

The procedure is the same irrespective of the size of article to be encapsulated. First ensure that it is quite dry, for the presence of moisture may retard the setting action and can also cause discolouration of the plastic. Mix by pouring your plastics into a mixing container and add the hardener, ensuring complete and even mixing by stirring well. Allow any bubbles that may have formed during the stirring to rise, and remove them. Ideally, you should aim at a gelling after say, five minutes and a complete hardening after fifteen minutes.

This will obviously be dependent upon the size of the article to be encapsulated, for large objects will take more time to fill and therefore a slower setting time should be allowed to provide time for the operation to be completed while the mix is still free flowing. Only mix sufficient for the job in hand, mixing fresh batches for each artifact, unless they are very small.

Place the object within the mould using any tools or devices that may be necessary to ensure that the object is correctly positioned and vertical, possibly even weighted if likely to float at all. Pour the mixture in very gently to prevent the formation of bubbles and be extremely careful not to allow any dust or extraneous matter to fall into the mould. Fill to just above the base of the object so that it is slightly immersed. The plastic will begin to gel after a time which will be dependent upon the amount and type of hardener or catalyst used, and the room temperature. While this is happening do be very careful not to knock or disturb the mould and contents and also try to maintain a constant temperature throughout the whole operation.

When the plastic has adequately gelled, it is sometimes possible to place the object in position, rather than doing it in the first instance. This can make the supporting devices rather less complicated than they would have to be to support the weight from the beginning. As soon as the resin has set sufficiently to hold the object without outside support, pour in the remainder of the plastic to cover completely and fill the mould. Do this carefully and before the lower layer has set completely and both gels will bond together with no sign of a joint. Even if it is left a little too long and a line is formed between both surfaces, this is only likely to be seen from the side with the eye level with the joint.

The objects may be left to cure in their own time and at room temperature, but accelerated curing in an oven at a temperature appropriate to the plastic and hardener chosen, will ensure better results. In deciding on the temperature, consider the object being encapsulated and be sure the temperature is not too high for its good. The supplier of resin and hardener will advise you on the correct curing temperatures.

When the finished encapsulation has cured (which process should not be hurried) it is very easy to cut and shape the solid block with ordinary tools such as saw and files, and finish off by polishing into a fine shiny surface. The secret of cutting is to take little and often, inspecting between practically every cutting stroke. The material is relatively soft and easily cut away to greater extent than required or desired. When finally cut to shape, bring to a perfect finish with a very fine wet-and-dry paper well lubricated with soap and water, followed by lots of elbow grease and metal or 'Perspex' polish. If rotating mops are used on a power polisher, set the speed at 2–3,000rpm and use very light pressure. Too much pressure or too little polish will cause overheating with the result that the acrylic will become burnt and discoloured. It pays to have very special specimens encapsulated professionally, for once they are embedded, it is not easy to get them out again for a second attempt! Professionally done, it will be quite a reasonably priced job and can be superbly carried out.

PUBLICATION

The second part of this chapter is concerned with reporting or publication and the only really important thing I would stress is that all work carried out should be reported, be it a large or small scale operation. This does not involve any great expense on the part of the finder, for it does not mean writing and publishing an enormous tome complete with plates and drawings etc. If only a typewritten and duplicated handout is prepared, with some drawings and perhaps photographs stuck in and is then circulated to interested parties, a tremendous amount will have been achieved.

To whom reports are circulated is really up to an individual, but I would stress that there is one group to which they must be sent, and that is to all those who have supported or sponsored the work in some way. All too frequently, I have heard the complaint that such and such a group was supported and nothing was heard from them. Not only is this, I think, gross discourtesy but it is damning to that group's chances, and probably those of anybody else, of getting further support on future occasions. Other than that, I would recommend that UK teams send their reports to the BS–AC, the CNA, the local archaeological society in the vicinity of the site, and the Royal

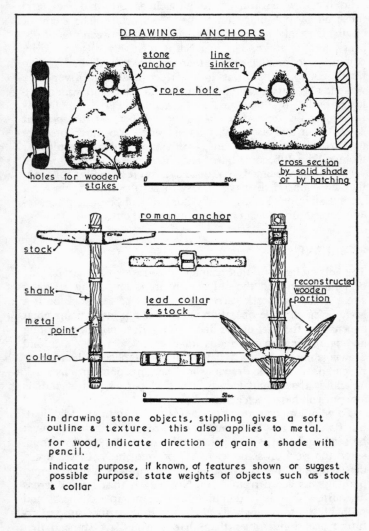

Geographical Society, the latter particularly if charting is involved in little-known areas.

Conservation and reporting

In America reports would probably be welcomed by the Smithsonian Institution and other naval/maritime museums, together with local archaeological societies and sponsoring organisations. In the Mediterranean areas it is also wise to send reports to the appropriate British School of Archaeology for the area, the national maritime and archaeological museums of the countries concerned, and the local authorities with which the team will have been working.

While the presentation may be simple, it should include certain specific headings amongst which should be:

Historical Background: Draw a pen-picture of the history of the site; if a wreck, how it came to be, if a submerged harbour or roadstead, who has used it, when, and why. Tell of previous investigations, when and by whom conducted, and what their findings were.

Expedition aims: Explain these; are they purely archaeological or also geological, ecological, etc?

Preliminary Research: Who has carried it out and where, what problems were experienced and how these were overcome. Contacts who have proved helpful, copies of interesting and relevant reports and letters. References used and where obtained.

Location: This is in respect of the site to adjoining land mass, means of access, local facilities. General description of the seabed in the area, the weather conditions, tides, currents, etc. Plans of the actual site and elevations of the same.

The Site: General description of the remains, whether scattered or relatively closely grouped. Description of work methods and patterns, tools used, devices conjured up to tackle particular problems, general diving conditions and problems. Sketch plans of the area, drawings and photographs of site and artifacts.

Analysis of finds: Methods of identification and dating, relation to other sites (if any), possible cause of wreck if not already

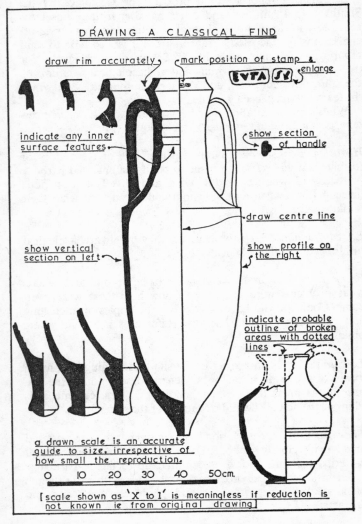

known. Notations on spread of jettisoned equipment, ie anchors, cannon, etc. Possible clues as to its previous journeys,

Conservation and reporting

gleaned from examination of cargo and artifacts. Detailed catalogue of finds, condition, and treatment of same.

Conclusions: Your interpretation of the site, etc.

The above are just an indication of the types of headings that one should endeavour to cover, they may well be amplified or deleted! Just try to set out a reasonable, factual report, logical interpretation and sensible conclusions in as simple a style as you wish, but please do it!

On the actual presentation of the contents, I would suggest that you examine other books or reports and model your own effort on the ones that impress you most. Photographs will enliven any report and should be included if at all possible. Concerning drawings: it is most important to do these well. Many potential readers will first skim through a book looking at the 'pictures'. If they are too awful, it is unlikely they will bother to go much further, so it will pay an author to consider first whether he knows someone who might be persuaded to do the drawings and make a better job of it than he could himself.

Simple line drawings carried out with drawing instruments are well within the capabilities of the majority of people and can provide a good basis for the work. But keep away from perspective drawings unless you really know what you are doing; they can be a veritable death-trap to inexpert artists or draughtsmen. Similarly, a good simple line drawing can be ruined by 'artistic' bits of seaweed, etc, so try not to embellish the subject, just report it in straight lines. Finally, figures of divers, hands, and heads. Keep away from them like the plague! How often does one see really excellent line drawings and plans completely ruined because the illustrator has stuck in a body? Unless you have artistic ability well above average, leave figures alone.

Even plans can be spoiled by bad lettering and numerals, and there is absolutely no excuse for this at all. No matter how bad the draughtsman's own lettering, it is easy enough to buy 'Letraset' or similar dry impact lettering from any art shop or commercial stationers and their use will add immensely to the presentation. The cost is negligible. For reproduction purposes,

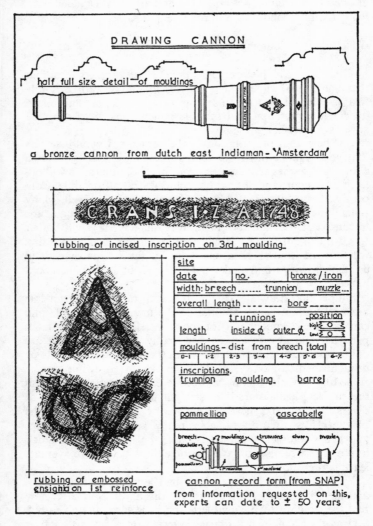

drawings will be reduced, perhaps to a third of their drawn size, so remember this and make sure all important lines and

Conservation and reporting

figures are very firm and clear. For preference, I would recommend plastic paper; this is easy to draw upon and first-rate prints can be taken off. Always show scale (metric please) and use acknowledged chart symbols where possible; these can be obtained from naval charts. Clearly indicate north and then title it, date it, and sign it—why not, it's your work!

If time and space permit, do not be afraid to include anecdotes relevant to the work. Many little happenings, unimportant in themselves, can collectively help to give a much better picture of the whole operation and make for more interesting reading. It is not necessary to be a great writer with tremendous journalistic ability to write an interesting and valuable report. But it does take a tremendous amount of work and time, probably a lot more than the actual expedition itself, and an ability to lay it out in a sensible order. Having completed your work, you may have mixed feelings as to the exact contribution you have made to this fast growing science. Your report will be, in essence, only a report of your own work and experience and you can but hope that in some small way it may be of assistance and interest to others.

Those are my feelings as I write this! No book on so wide a subject can be complete in its coverage. In many areas I have skimmed over the surface, referring to others who are more competent than myself in those fields. I can only speak from my own experience and learnings from some twenty years of diving and eight years of underwater archaeology.

I have tried to treat everything fairly simply and if I have overdone this, it is out of no disrespect for the reader's abilities, but as a counter to the mystique that some seem to feel so necessary in describing their own work. Many of the suggestions I have outlined will be well known to some divers, some of whom may have improvements to suggest—but if my thoughts here help even a few to obtain half of the fun, satisfaction, and friends that I have been fortunate enough to gain in the course of this work—then that will be reward enough.

BIBLIOGRAPHY

THE LIST of publications concerned with underwater archaeology increases almost daily, and with the development of technology in this field more and more technical papers are being produced. The compilation of a complete and up-to-date bibliography is well beyond the scope of this book.

I have therefore tried to include a selection of those publications that will help newcomers to the techniques of underwater archaeology and have omitted those books that relate to diving alone or which concentrate on the case histories of specific underwater work.

I understand that Gerhard Kapitan has compiled a bibliography running into thousands of titles, which is to be published by the Council of Underwater Archaeology in San Francisco. This should provide a most valuable source of references.

GENERAL

Bass, G. F. 'The Cape Gelidonya Wreck', *American Journal of Archaeology*, 65 no 3, (1961) 267–76

Bass, G. F. 'Underwater Excavation at Yassi Ada', *Archaeologischer Anzeiger*, (1962) 537–64

Bass, G. F. 'The Asherah, a submarine for archaeology', *Archaeology*, 18 no 1 (1965) 7–15

Bass, G. F. *Archaeology Under Water*, New York, 1966

Bass, G. F. & Throckmorton, P. 'Excavating a Bronze

Age shipwreck', *Archaeology*, 14 no 2 (1961) 78-88

Broadribb, C. *Drawing Archaeological Finds*, 1970

Cousteau, J-Y. & Dumas, F. *The Silent World*, New York, 1953

Dumas, F. *Deep Water Archaeology*, 1962

Falcon-Barker, F. *Roman Galley Beneath the Sea*, Philadelphia, 1964

Flemming, N. 'Underwater Adventure in Appolonia', *Geographical Magazine*, 31 (1959) 497; 'Appolonia revisited' ibid, 33 (1961) 522

Franzen, A. *The Warship Vasa*, Stockholm, 1960

Frost, Honor, *The Mortar Wreck at Mellieha Bay*, 1970

Frost, Honor, *Under the Mediterranean*, 1963

Grimsell, Rahtz & Warhurst *The Preparation of Archaeological reports*, 1970

Hampton, T. A. *The Master Diver*, Southampton, 1955. Revised edition, Newton Abbot, 1970

Hume, I. N. *A Guide to Artifacts of Colonial America*, New York, 1970

Jensen, C. W. & J. *Modern Laws Relating to Salvage and Sunken Treasure*, Natick, Mass, 1960

Johnston, C. S., Morrison, I. A. and MacLachlan, K. 'A photographic method for recording the underwater distribution of marine benthic organisms', *Journal of Ecology*, 57 (1969) 453-59. An illustration of the box-frame technique

Kenyon, L. *The Pocket Guide to the Undersea World*, New York, 1956

Link, Edwin A. 'Special equipment for underwater archaeology'. Paper presented at London Conference on Underwater Activities, 1962

Link, M. C. *Sea Diver, A Quest for History under the Sea*, 1959

McKee, A. *History Under the Sea*, 1969

Morrison, I. 'An inexpensive photogrammetric approach to the reduction of survey diving time', *Underwater Association Report* (1969) 22-8

Ohrelius, Bengt. *Vasa, The King's Ship*, Philadelphia & New York, 1963

Olsen, S. J. 'Scuba as an aid to underwater archaeologists and paleontologists.' *Curator*, 4 no 4 (1961)

Petersen, M. *History Under the Sea*, Washington, 1965
Ryan, E. J. & Bass, G. F. 'Underwater surveying and draughting', *Antiquity*, 36 no 144 (1962)
Taylor, J. du Plat. ed. *Marine Archaeology*, 1965
Throckmorton, P. *Shipwrecks and Archaeology*, 1970
Throckmorton, P. & Bullitt, J. 'Underwater surveys in Greece'. *Expedition*, 5 no 2 (1963) 16–23
Throckmorton, P., Hall, E. T., Frost, Honor, Martin, C., Walton, M. C. & Wignall S., *Surveying in Archaeology underwater*, 1969
Woods J. D. 'The role of underwater science', *Underwater Association Report* (1968), 1–2

SEARCH AND SURVEY

Alldred, J. C. 'The fluxgate gradiometer for archaeological surveying', *Archaeometry*, 7 (1966)
Atkins, M. J. 'Magnetic prospecting', *Archaeometry*, 1 (1958)
Atkins, M. J. *Physics and Archaeology*, 1961
Bass, G. F. & Katzev, M. L. 'New tools for underwater archaeology', *Archaeology*, 21 (1968)
Block, Hansen & Packard, *Physics Review*, 70 (1946)
Colani, C. 'A new type of locating device', *Archaeometry*, 9 (1966)
Daniels, C. & Henderson, R. 'An integrated underwater survey system'. Paper given at Oceanology International, Brighton, 1969
Edgerton, H. E. & Payson, H. 'Sediment penetration with a short pulse sonar'. Paper at the 44th Annual Meeting, American Geophysical Union, 1963
Edgerton, H. E. & Yules, J. 'Bottom Sonar search techniques'. *Undersea Technology*, 5 no 11 (1964)
Edgerton, H. E. *Sonar for Sub-bottom Penetration*, Monaco, 1968
Foster, E. J. 'Further developments of the pulsed induction metal detector', *Prospezioni Archaeologiche*, 3 (1968)
Green, J. N., Hall, E. T. & Katzev, M. L. 'Survey of a Greek Shipwreck off Kyrenia, Cyprus', *Archaeometry*, 10 (1967)
Green, J. N. *Cape Andrea Expedition, 1969*, Oxford, 1969

Hall, E. T. 'Use of a Proton Magnetometer in Underwater Archaeology', *Archaeometry*, 9 (1966)

Harnett, C. B. *Lodestone Operation*, Orlando, Fla, 1962

Henderson, R. *Application Notes on Side Scan Sonar*, 1968

Klein, M. 'Side Scan Sonar'. *Undersea Technology*, April 1967

Knott, S. T. & Hersey, J. B. 'Interpretations of High Resolution echo sounding and their use in bathymetry', paper given at Oceanology International, Brighton, 1969

Moody, D. W. & Van Reenan, E. D. 'High Resolution sub-bottom profiles of the Delaware Estuary'. Paper at the 44th Annual Meeting, American Geophysical Union, 1963

Van Reenan, E. D. 'Sub-surface exploration by sonar seismic systems', *Telephony* (1964)

Van Reenan, E. D. & Smith, D. A. 'Hydrographic and seismic profiling surveys', *Telephony* (1966)

Williams, J. C. C. *Simple Photogrammetry*, 1970

Waiatr, Z. M. & Parsons, L. W. 'Rubidium Vapour Magnetometer', *J Scientific Instruments*, 39 (1962)

CONSERVATION

Albright, A. 'The preservation of small waterlogged wood specimens with polyethylene Glycol', *Curator*, 9 no 3

Brill, R. H. 'The record of time in weathered glass', *Archaeology*, 14 no 1 (1961)

Ciba Co Inc. 'A method of wood preservation using arigal WP'. Provisional circular, 1961

Dowman, E. A. *Conservation in the Field*, 1970

Erickson, E. & Thegal, S. *Conservation of Iron Recovered from the Sea*, Copenhagen, 1966

Gettens, R. J. 'Mineral alteration products on ancient metal objects', in Thomson, G. *Recent Advances in Conservation*, 1963

Gettens, R. J. & Usilton, B. M. 'Abstracts of technical studies in art and archaeology, 1943-52', *Freer Gallery of Art, Occasional Papers*, 2 no 2

Hodges, H. W. M. *The treatment of iron recovered from the sea*. Preliminary paper, Institute of Archaeology, 1970

Katzev, M. L. & van Doorninck, F. H. 'Replicas of iron tools from a Byzantine shipwreck', *Conservation*, 11 no 3

Muller-beck, H. & Haas, A. 'A method of wood preservation using arigal C', *Studies in Conservation*, 5 (1960)

Organ, R. M. 'The conservation of fragile metal objects', *Conservation*, 6 no 4

Pittsburgh Plate Glass Company. *Embedding Objects in Selectron 5000 Resin*, Pittsburgh, Pa. 1946

Plenderleith, H. J. *The conservation of antiques and works of art*, 1956

Seborg, R. M. & Inverarity, R. B. 'Conservation of 200-year-old waterlogged boats with polyethylene glycol', *Studies in Conservation*, 7 (1962)

Smith, J. B. & Ellis, J. P. 'The preservation of underwater archaeological specimens in plastic', *Curator*, 6 no 1 (1963)

Werner, A. E. 'Consolidation of fragile objects', *Conservation*, 6 no 4 (1961)

IDENTIFICATION—ANCHORS AND ARTILLERY

Cloves, W. L. *The Royal Navy*, Boston, 1898

Cotsell, G. *A Treatise on Ship's Anchors*, 1856

Ffoulkes, C. *The Gun Founders of England*, Cambridge. 1937

Frost, H. 'From Rope to Chains', *Mariners Mirror*, 49 (1965)

Gargallo, P. N. 'Anchors of antiquity', *Archaeology*, 14 (1961)

Grant, M. *Armada Guns*, 1961

Manucy, A. *Artillery Through the Ages*, Washington, DC, 1949

Moll, F. 'The history of the anchor', *Mariners Mirror*, 13 (1927)

Muller, J. *A Treatise of Artillery*, 3rd ed, 1780

COINS

Adams, E. H. *Catalogue of the Collection of Julius Guttag (Latin American Coins)*, New York, 1929

Brooke, G. C. *The English Coins from the Seventh Century to the Present Day*, 1932

Peck, C. W. *English Copper, Tin and Bronze Coins in the British Museum, 1558–1958*

Seaby, B. A. *Standard Catalogue of the Coins of Great Britain & Ireland*, 1929

Seaby, H. A. & Seaby, P. J. *British Copper Coins and their Values*, 1949

Santos, Leitao & Cie. *Catalogo de monedas Brasileiras de 1643–1844*

Wood, H. *The Coinage of the West Indies, and the Sou Marque*, New York, 1915

Yriarte, J. de. *Catalog de los Reales de a Ocho Espanoles*, Madrid, 1955

POTTERY

Bemrose, G. *Nineteenth Century English Pottery and Porcelain*, 1952

Benoit, F. 'Amphores grecques d'origins ou de provenance Marseilles', *Rivista di Studi Liguri*, 21 (1955)

Benoit, F. 'Amphores et ceramiques de l'epave de Marseilles, *Gallia*, 12 (1954)

Benoit, F. 'Typologie des amphores', *Gallia*, 14 (1956)

Cushion, J. P. & Honey, W. B. *Handbook of Pottery and Porcelain Marks*, 1956

Dressel. 'Amphora types', *Corpus Inscriptionum Latinorum*, 15 (1899)

Edwards, G. R. 'Hellenistic pottery from the Antikythera wreck', *American Journal of Archaeology*, 64 (1960)

Goggin, J. M. *The Spanish Olive Jar, An Introductory Study*, New Haven, 1960

Grace, V. R. *Amphora and the Ancient Wine Trade*, 1961

Rackham, B. & Read, H. *English Pottery, its Development from Early Times to the End of the 18th Century*, 1924

Thorn, C. J. *Handbook of Old Pottery and Porcelain Marks*, New York, 1947

Watkins, C. M. 'North Devon Pottery and its export to America in the 17th century', *United States National Museum Bulletin*, 225

Webster, C. *Roman-British Coarse pottery*

GLASS

Haynes, C. B. *Glass through the ages*

Hume, I. N. 'Dating English glass wine bottles', *Wine & Spirit Trade Record*, Feb 1955

Hume, I. N. 'The glass wine bottle in Colonial Virginia', *Journal of Glass Studies*, 3 (1961)

McKearin, G. S. & McKearin, H. *American Glass*, New York, 1941

Ruggles-Brise, S. M. E. 'Sealed Bottles', *Country Life*, New York, 1965

Wills, G. 'Bottles', *Guinness Signatory Series*, 9 and 10, 1970

PIPES

Hall, T. M. *On Barum Tobacco-pipes and North Devon Clays*, Plymouth, 1890

Harrington, J. C. 'Dating stem fragments of seventeenth century clay tobacco-pipes', *Quarterly Bulletin, Archaeological Society of Virginia*, 9 no 1 (1954)

Hartley, L. S. *The Clay Tobacco Pipe in Britain*

Helbars, G. C. & Goedswaagen, D. A. *Pipes de Gouda*, Amsterdam, 1942

Oswald, A. H. 'English Trade tobacco pipes', *Archaeological News Letter*, 3 no 10 (1951)

Oswald, A. H. 'The archaeological and economic history of the English clay tobacco pipe', *Journal of British Archaeological Association*, 23

Price, F. G. H. 'Notes upon some early clay pipes from the sixteenth to the eighteenth century', *Archaeological Journal*, 57 (1900)

Walter, I. C. 'Clay pipes from the fortress of Louisberg', *Archaeology*, 20 no 3. Discusses Dutch pipes.

APPENDIXES

1 POSSIBLE SOURCES OF INFORMATION CONCERNING WRECKS: UK

Museums

The Director, British Museum, London WC1

J. Horsley, Curator, Brixham Museum, 18 Sellick Avenue, Brixham, Devon.

British Archaeological Research Group, City Museum, Queens Road, Bristol 8.

T. Henderson, FSAScot, County Museum, Lower Millhead, Lerwick, Shetland Isles.

The Curator, Cumberland House Museum, Portsmouth, Hants.

P. R. V. Marsden, Guildhall Museum, London EC3.

J. Mainwaring-Baines, Curator, Hastings Museum, Hastings, Sussex.

E. W. Paget-Tomlinson, Keeper of Shipping, City of Liverpool Museum, Liverpool.

Maritime Museum, Bucklers Hard, Beaulieu, Brockenhurst, Hants.

The Director, National Maritime Museum, Greenwich, London SE10

R. B. K. Stevenson, National Museum of Antiquities, Queen Street, Edinburgh 2.

Mr Chaff, Curator, Naval Museum, HM Dockyard, Plymouth, Devon.

The Curator, Plymouth City Museum, Tavistock Road, Plymouth.

The Director, Science Museum, South Kensington, London SW7.

Miss S. Butcher, Scillies Museum, St Mary's, Isles of Scilly.

The Curator, Southampton Museum, St Michael's Square, Southampton, Hants.

A. N. Kennard, Assistant Master of the Armouries, HM Tower of London, London EC3.

The Director, Ulster Museum, Belfast, Northern Ireland.

The Curator, Victory Museum, HM Dockyard, Portsmouth, Hants.

Government or service departments

A. C. Carpenter, Superintendent, Ancient Monuments Branch, Ministry of Public Buildings and Works, Pearl Assurance House, Royal Parade, Plymouth.

Lt Cdr M. Godfrey, RN (retired), Admiralty Records, Public Records Office, Chancery Lane, London WC2.

The Librarian, BOT Marine Library, Room 82, Norman Shaw Building (N), Victoria Embankment, London SW1.

The Secretary, Crown Estate Commissioners, Crown Estate Office, Whitehall, London SW1.

Dr P. N. Davies, Department of Economics and Commerce, Social Services Building, Bedford Street South, Liverpool 7.

Sir Patrick Kingsley, KCVO, Duchy of Cornwall, 10 Buckingham Gate, London SW1.

The Librarian, Head Office, HM Customs and Excise, King's Beam House, Mark Lane, London EC3.

The Curator, Hydrographic Department, MOD, Taunton, Somerset.

Director General Defence Contracts, (CP 85), MOD, Bath BA1 5AD, Somerset.

Deputy Under Secretary (Navy), Room 4177, MOD, Main Building, Whitehall, London SW1.

The Librarian, Navy History Library, Plymouth, Devon.

The Naval Library, Empress State Building, Fulham, London.

The Officer In Charge, Royal Naval Lifeboat Institute, Operation and Coast Personnel Department, Lifeboat House, 42 Grosvenor Gardens, London SW1.

The Officer in Charge, Wreck Section, MOD, Taunton, Somerset.

Appendixes

Others

Prof C. R. Boxer, Ringshall End, Little Gaddesden, Berkhamsted, Herts.

The Secretary, British Ship Research Association, Prince Consort House, 27-29 Albert Embankment, London SE11.

K. R. Mason, Shipping Editor, Corporation of Lloyds, Lime Street, London EC3.

T. Griffiths, The Warden, Dartington Hall, Dartington, Near Totnes, Devon.

R. Davis, 3 Berrycombe Hill, Bodmin, Cornwall.

F. S. Dunn, Swn-y-Mor, Lizard, Helston, Cornwall.

G. Farr, D. Bradford Barton, Ltd, Truro, Cornwall.

R. Faux, 4 Byrneside, Hildenborough, Kent.

A. Flinder, Aldwych House, Aldwych, London WC2.

Dr E. T. Hall, 6 Keble Road, Oxford.

The Managing Director, Imray Norrie & Wilson, St Ives, Huntingdon.

Mrs M. Rule, Secretary, Joint Archaeological Committee, Mill House, Westbourne, Hants.

R. Larn, 3 Barton Close, Gwealdues, Helston, Cornwall.

J. D. Macauley, 102 Frederick Crescent, Port Ellen, Isle of Islay, Argyll.

Merchant Shipping and Record Office, General Register of Seamen and Shipping, Llantrisant Road, Llandaff, Cardiff.

J. Owen, 30 Lucas Lane, Plympton, Plymouth, Devon.

The Worshipful Company of Shipwrights, Baltic Exchange Chambers, 24 St Mary Axe, London EC3.

Siebe Gorman & Co Ltd, Records Dept, Neptune Works, Davis Road, Chessington, Surrey.

Cpl G. Steward, RAF Station Halton, Nr Aylesbury, Bucks.

The Corporation of Trinity House, Trinity House, Tower Hill, London EC3.

R. Craig, BSc (Econ), FRHistS, Lecturer in Economic History, University College, London WC1.

A. Binns, BA, Department of English, University of Hull, Hull.

M. Vinnicombe, Drusilla Cottage, Halvasso, Mabe, Near Penryn, Cornwall.

W. St John Wilkes, 47 Pashley Road, Eastbourne, Sussex.

2 POSSIBLE SOURCES OF INFORMATION CONCERNING WRECKS: OVERSEAS

Australia. National Library of Australia, Processing Branch. Canberra ACT; Western Australia Museum, Perth.

Belgium. National Scheepvaartmuseum, Steenplein, 1, Antwerp.

Canada. W. Buck, Royal Canadian Air Force, Lazo, BC.
Maritime Museum, The Foot of Cypress Street, Vancouver 9, BC.
Maritime Museum of British Columbia, 28 Bastion Square, Victoria, BC.

Denmark. Danish Admiralty Collection, Naval Museum, Royal Dockyard, Copenhagen.
Ole Crumlin-Pedersen, Curator of Ships, Danish National Museum, Vikingeskibshallen, Roskilde.

France. Underwater Archaeology Commission, CMAS Secretariat, 34 Rue du Colisee, Paris.

Finland. Bureau of Maritime Archaeology, National Museum, Helsinki.
Neptune Club Skindivers of Turku, (Abo).
Murena Club of Porvoo, (Borga).

Netherlands. Mrs M. A. P. Meilink-roelofsz, Keeper of the First Section, Algemeen Rijksarchief, s-Gravenhage, Bleijenburg, 7.
Nederlandsch Historisch Scheepvaart Museum, De Lairesse-Hoek Corn, Schuystraat 57, Amsterdam.
Maritime Museum Prins Hendrik, Burg S'Jacobplein 8, Rotterdam.
G. D. Van de Heide, Museum Schokland, Middlebrund Schokland, post Emmelbord.
Rijksmuseum Library, Rijksmuseum, Amsterdam.

Norway. Svein Molaug, Museumdirektor, Norst Sjofartsmuseum, Bygdones, Oslo.

New Zealand. Wade Doak, New Zealand Maritime Museum, PO Box 20, Whangeri.

Portugal. Museo de Marinha, Lisbon.

Sardinia. Pro F. Barrecca, Museo Nationale, Cagliari.

Spain. Museo Naval, Madrid.

Sweden. The Director, Sjohistoriska Museet, Stockholm.
USA. Edward P. Von der Porten, 2250 Grahn Drive, Santa Rosa, California 95404
George F. Bass, University of Pennsylvania, Pa.
The Director, Museum of History and Technology, Smithsonian Institute, Washington, DC.
Mariners Museum, Newport News, Virginia, 23606.
Peabody Marine Museum, Salem, Mass.
US Naval Academy, Annapolis, Md.

3 CONSERVATION

Listed below are museums in the UK whch have conservation facilities. In general, the conservation officers attached to them will be prepared to offer advice on treatment of artifacts, but rarely will they be able to undertake work on conservation unless there are very special circumstances. Although not listed below, the major museums in other countries can usually be approached for advice, but again it is unlikely they will be able to undertake any actual work.

Birmingham. Conservation Assistant, City Museum and Art Gallery, Congreve Street, Birmingham 3.
Bristol. Conservation Assistant, City Museum, Queen's Road, Bristol 8.
Edinburgh. Research Laboratory, National Museum of Scotland, 5–6 Randolph Crescent, Edinburgh 3.
Liverpool. Conservation Dept, City Museum, William Brown Street, Liverpool, 13 8EN.
London. Conservation Dept, Institute of Archaeology, 31–34 Gordon Square, London WC1.
Oxford. Conservation Assistant, Oxford City & County Museum, Fletcher's House, Woodstock, OX7 1SN.
Southampton. Conservation Laboratory, Dept of Archaeology, The University, Southampton.

4 THE LAW OVERSEAS

IN CHAPTER 3, it was shown that British salvage laws did not really cover ancient wrecks of possible archaeological value, allowing salvors to excavate without regard for important historical or archaeological considerations. As all such wrecks could contribute to our knowledge of shipbuilding and of the everyday equipment of their time, they constitute an extremely valuable but numerically limited resource.

It is hoped that amendments to British salvage laws currently being considered in Parliament will close at least some of the loopholes. But it is disturbing to see that in October 1970, a judgement was passed in a Scilly Isles court to the effect that a pair of navigational dividers recovered from the *Association*, were personal property and as such did not even come under the current laws of salvage! Presumably, it could be interpreted that any coin, plate or jewellery found on a wreck site could be claimed to have been a personal possession, and could be free to be taken at will. Such a situation would be disastrous to archaeology and we can only hope that the proposed Antiquities Bill will cover this point especially. It is really tragic that, of all countries, Britain alone should have no laws covering this subject.

Over forty years ago the law of Northern Ireland was altered to provide for the reporting and protection of all finds of archaeological objects. This was done by an amendment of the Ancient Monuments Act, which up to that time, was common to Britain and Ireland. So Northern Ireland effectively protects her historical heritage. During the 1940s, equivalent laws were enacted in Eire and the Isle of Man, and as far back as 1904, an act was passed in India, then part of the British Empire, which protected antiquities. Other colonial territories provided laws on this subject, right through the twenties, thirties, forties and fifties—only England and Scotland have remained unprotected by an antiquities statute.

Elsewhere in the world, most countries have equipped themselves with laws of this kind. The League of Nations and United Nations Conferences have recognised and encouraged the passing of statutes by various countries aimed at the protection of antiquities. So, summing up, outside of England and Scotland, it is necessary to study the laws of the country around which you intend to dive and comply carefully with their requirements.

Most Mediterranean countries lay claim to all and any wreck

of historical, ancient or intrinsic value as their own and adopt stringent measures to ensure that no unauthorised exploration takes place. When permission is granted to allow exploration and excavation, it is usually insisted that a qualified, and sometimes local, archaeologist is present at all times. He need not be a diver, of course, but will effectively supervise all work being carried out and will ensure that accurate records are maintained of the work done and finds brought up. Artifacts salvaged must be handed to the local or appropriate museum and there is no certainty that any award will be made to the salvor, although more often than not some recompense is allowed in kind if not in cash. Frequently, the laws governing underwater archaeology are merely extensions of those applicable to land work. It is not sufficient to obtain verbal or even written approval from a local museum or archaeologist before starting work. In 1969, a UK team obtained written permission to dive in Yugoslavia from the local people, but on arrival were absolutely forbidden, even though they promised not to raise any artifacts. The requisite permission should have been obtained from the Ministry responsible for antiquities. A regulation passed on 1 January 1970 requires a written permit for any diving with compressed air, also a permit is required for navigating an inflatable boat.

Greece generally does not discourage skin divers, but compressed air must not be used. It is expressly forbidden unless written permission is obtained and a copy of the permit carried at the site. The penalties for infringement range from a minimum of the confiscation of equipment to quite long terms in prison. Even where permission has been given to dive on an archaeological site using compressed air, it is conditional upon no excavation or lifting of artifacts being carried out. In very special circumstances, permission may be obtained to excavate a site, but only under close supervision and with all finds being returned to a museum.

In Italian waters, permission must be obtained from the Soprintendenza all Antichita of the province in which you wish to work, from the Ministero della Pubblica Istruzione Direzione Generale delle Antichita e Belle Arti and should be backed up by a recommendation from the British School at Rome. This permit, or copy, should be available for inspection at your base camp at all times. In general, with those countries in which there is a British School of Archaeology, it is important to have their backing.

In the United States it appears that there are no Federal laws covering underwater archaeology, wrecks or salvage of ancient materials. But in common with other major powers, the international laws of salvage apply—though as we have explained, these do little to help. It has been left to individual states to make their own ruling. The two foremost in this respect are Florida and Texas, not unnaturally in view of the wealth of underwater wreck, mostly dating back to the Spanish treasure fleets, lying in their coastal waters.

Taking Florida first, the Florida Archives and History Act of 1967 was introduced and in it all aspects of archaeology were covered. Section 267.061 applies to

> all articles of ancient, historic or intrinsic value, recovered on or beneath land, where title is vested in the State of Florida or any state agency and all state-owned sovereignty submerged lands

Under Section 41B–1.04, all such articles that have been abandoned on land as set out above, are declared to belong to the State of Florida. Any unauthorised recovery of these articles is punishable on conviction.

Under this Act, a Florida Board of Archives and History was established and consists of the Governor, the Secretary of State, the Attorney General, the Comptroller, the State Treasurer, the Superintendent of Public Instruction and the Commissioner of Agriculture. A higher level of authority cannot be imagined. The board has an advisory council consisting of the Director of the Florida State Museum, the State Geologist, representatives from all educational institutions offering graduate degrees in anthropology or archaeology, the presidents of the Florida Historical Society, Florida Anthropological Society, and the Florida Library Association. The directors of the Division of Recreation and Parks, of the Board of Trustees of the Internal Trust Fund and of the university presses of any state university. The actual day by day administration is vested in a director appointed to the Division

Appendixes

of Archives, History and Records Management who supervises, directs and co-ordinates the activities of the division and its various bureaux.

With such a powerful administrative and advisory organisation, very tight control can be effected over all activities involving above or under water work. No work may be carried out in either medium without a proper contract issued by the board, and the contracts provide for, initially, exploration and then excavation.

The 'contract for exploration' requires a payment to the board of $600 as annual rental, and permits the holder to explore a precisely defined area, for discovery purposes only, in search of

> salvageable abandoned vessels or remains thereof, relics, treasure trove and other articles and materials contained in or on the submerged lands aforesaid; said area being State owned submerged lands in or on which abandoned personal property is acknowledged by the parties hereto to be the property of the State of Florida where found; all subject to the rights of the upland owners adjacent to such submerged lands and above the mean high water mark.

The contract is issued subject to certain covenants and the explorer must lodge a bond of $5,000 to guarantee that he will abide by them. The covenants cover the requirement to report at least at three-monthly intervals, and that a log be kept in a logbook supplied by the board and which remains the property of the board. Both log, accounts and books may be audited by the board at any time. No objects may be lifted unless in the presence of an agent of the board and no explosives may be used. He is required, immediately upon location, to advise the board of the precise position of any salvageable materials.

If he does this to the satisfaction of the board and the location of salvageable material is confirmed by the department; he may then be granted a 'contract for salvage'. This is similarly tied with covenants and subject to an annual rental of $1,200. In the first instance, only one year is agreed but this may be renewed if all is satisfactory. With this contract a bond of $15,000 must be lodged to guarantee compliance with the covenants and, again, the use of explosives is expressly forbidden and three-monthly reports, written under oath, must be submitted to the department. All boats used in the operation must fly a flag supplied by the department and prominently display a number designated by the executive director of the board. During all salvage operations an agent of the board must be present and no work may be done in his absence.

These and many other requirements are clearly aimed at the utmost control of divers and salvage operators and are so restrictive that only serious operators will enter into agreement to explore or salvage. However, the rewards to the successful salvor are higher than in the UK, for the department requires only 25 per cent of the value of the recovered goods, or 25 per cent of the items themselves, or a combination of both at the option of the department. In any event the salvor stands to gain 75 per cent of his finds.

This fairly generous basis of settlement, coupled with the fact that each settlement is settled as expeditiously as possible after the reporting of the finds, tends to encourage responsible activities. The penalties for exploring, lifting or salvaging without contract very much tend to discourage such unauthorised activities.

It has been found that, even with these tight controls, insufficient attention has been given to the full archaeological study of the wrecks themselves, as separate from their contents, and existing conservation and preservation facilities have already been swamped with artifacts in need of attention. It was felt that this situation, if allowed to go unchecked would quickly deplete the valuable and limited number of wrecks off the Florida coastline.

It was therefore decided that specific areas containing a broad sample of historically significant wrecks should be set aside as reserve areas in which no contracts for exploration or salvage would be granted. Four such areas have been established and are clearly defined by geographical and chart references, which may be obtained from the department.

The seas off Texas also hold a wealth of interesting wrecks, and control of salvage was deemed necessary by that state. So an Antiquities Committee was created by the Antiquities Code of Texas, Senate Bill No 58, Acts of the 61st Legislature,

Appendixes

Second called Session, 1969 and codified as Article 6145-9 in Vernon's Civil Statutes.

The Committee's responsibilities cover all forms of archaeology, both on land and under water and one of its main functions is to designate as State Archaeological Landmarks any site of historic, archaeological or scientific interest. Under Section 5, this is extended to cover

> All sunken or abandoned pre-twentieth century ships and wrecks of the sea or any part or the contents thereof and all treasure imbedded in the earth, located in, on or under the surface of the lands belonging to the State of Texas, including its tidelands, submerged lands and beds of its rivers and the sea within the jurisdiction of the State of Texas are hereby declared to be State Archaeological Landmarks and are the sole property of the State of Texas and may not be taken, altered, damaged, destroyed, salvaged or excavated without a contract or permit of the Antiquities Committee.

Such contracts are only issued to those organisations or bodies that employ professional archaeologists in direct charge of the project and can provide proof that adequate funds, equipment and facilities are available to conduct the investigation and to publish the results, ten free copies of which must be furnished to the committee. Salvors are also responsible for the treatment and conservation of all finds, at their own expense.

Permits or contracts are only valid for one year, but may be extended if the work has been 'diligently prosecuted under the permit'. Separate permits are issued for 'survey and reconnaissance', which only allows for visual examination, plus use of magnetometers and metal detectors, mapping and photography, and some controlled surface collecting. If this has been carried out to the satisfaction of the committee a 'testing' permit is issued which allows detailed examination of the site plus test excavation and, finally, an 'excavation' permit will be issued which permits full investigation and extensive excavation. In each case the permit must be available on site for inspection at all times.

The Texas contracts allow for 'fair compensation' to salvor

in terms of a percentage of the reasonable cash value of the objects recovered or a fair share of the objects. What a 'fair share' is, is not specified, but in assessing it account is taken of the circumstances of the operation, with, if necessary an appraisal by qualified experts. So, while the rewards may not be so clearly stated in Texas as in Florida, the penalties are! These range from a fine of $50 to $1,000, or a confinement in jail for not more than thirty days or both. Each day of continued violation is treated as a separate offence for which the offender may be punished.

It is clear from the foregoing that these two states have clearly recognised the need to treat antiquities on land and under water as being of equal value and have extended their land controls to cover underwater activities. It does seem, however, that even here the emphasis is on the controlled recovery of the artifacts and not too firm a ruling is laid down concerning the recovery of wreck structure or even the detailed examination of such. Section 4(i) of the Salvage contract of Florida, does state

> Shall extend all reasonable cooperation to any archaeologist provided by the Department to prepare maps, make site studies, and salvage artifacts in the best interests of the State.

However, it is probably unlikely that the state archaeologist is a diver and could carry out the survey of an underwater site, so this responsibility falls back on the divers and we can only hope that they accept it and do carry out the necessary site/wreck survey as they recover the artifacts.

Even allowing for this possible criticism, these two states have set a standard that could well be followed by others and even other countries. It is interesting to note that the Texas Act under section 23 of the Antiquities Code states that

> The fact that irreparable damage ... creates an emergency and imperative public necessity that the Constitutional Rule requiring bills to be read on three several days in each House be suspended, and this Rule is hereby suspended; and this Act shall take effect and be in force from and after the date of its passage, and it is so enacted.

Appendixes

Space does not permit a detailed discussion of the laws of other American states. Application should be made to the appropriate state museums or archaeological societies, some of which are listed in Appendix 2.

INDEX

Acrylonitrile Butadiene Styrene (ABS), 80-1
'A-frames', for lifting, 226-7
Air lance, 117-18
Air lift, 208-17; air requirements, 214
Air to water interface, 91
Albenga wreck, 161
Amphora, 26, 27, 30, 58, 136
Amsterdam, 20-5, 41, 46, 51, 73, 122, 158, 161, 240, 243, 247, 260
Ancidonia, 33
Anchors, 30, 41, 129-31, 135, 241-2; for airlift, 210
Anchorages, 30, 139
Ancient Monument, 46
Apollonia, 31
Aqua level, 172-5
Archaeology, underwater, 13, 16, 18, 26, 62, 159-60
Artifacts, 48, 64, 133, 158, 169; lifting of, 226-37; drawing, 266-8
Association, 15, 41

Bags, lifting, 50
Ballast stones, 46, 133
Base lines, 51-2, 114, 150, 163-4; secondary, 166
Beaches, underwater, 32
Bearing Circle, 87, 171
Bithia, Sardinia, 242
Board of Trade, 36
Boat, safety, 97-8
Bottles, identification from, 247; markers, 56

Boomer, *see* sonar, sub-bottom
Buoyancy, 51, 118-19; devices, 231-4; capacities, 234
Buoying, 47, 62, 155, 157

Cannon, identification from, 238-41; -balls, 132; lifting of, 229, 233; sizes, 239
Cape Gelidonya, 161
Cargo mounds, 130
Causeways, 32
Ceramics, 257
Charts, Admiralty, 36, 46, 95, 151
Chemical changes in metals, 251
Chisels, use of, 205
Churches, 41
Clues, 95, 129-39
Coastguards, 45
Coffer dams, 159
Coins, dating from, 244
Colliers Letters, 22
Committee for Nautical Archaeology, 14-16, 23, 26, 36, 40, 45, 48
Compass, use of, 51, 55, 78, 85, 99, 102, 112, 114, 143; compass, theodolite, 156
Communication underwater, 99
Concretion, calcareous matter, 132, 250-1
Conservation, 248-50; containers, 250; chemicals, 250-2; ceramics, 257; organic objects, 257; leather, 258; wood, 258-9; glass, 260
Consuls, foreign, 48

Copper cladding, 243
Cork, 56
Corers, 115-18
Counterweight buoying, 62
Crannogs, 15
Crosses, photogrammatic, 197
Crowbars, 205
Crown, rights of, 42
Currents, 46, 134-5, 139; barriers, 206-7
Customs records, 39

Datum points, 88-90, 170, 173
De Liefde, 14, 72
Depth gauge, 55, 88
Derelict, 42
Displacement values, 236
Diver floats, 103
Dowels, wooden, 243
Drawing equipment, 78, 178; Artifacts, 178, 267-9
Dredges, 221-4
Dressel, 27
Droits of Admiralty, 42
Dutch East India Company, 14, 20

Earthquakes, 32, 137
Echo Sounders, 88, 123-5
Elaphonisos, 183
Electrochemical treatment, 254
Electrolytic action, 132-3; treatment, 255
Embolism, 107
Encapsulation, 261-3
Entrenching tools, 204
Epoxy resins, 70
Excavation, 202-6; sand, 224
Explosives, 225

Fastenings, 227-9, 242-4
Finds Book, 177
Flotsam, 42

Geology, 133, 137
Gerona, 41
Grace Dieu, 15
Grand Congloue, 161, 210

Grids, 64, 67-8, 114-15, 174-7; photographic, 183-4
Growth rings, 28

Harbour Master, 45; works, 48, 135, 138, 161
Hydrographic Office, 41, 46

Ice-tongs, 230
Identification, 39
Inclinometers, 174, 194-5
Ireland, Republic of, 43
Iron, chemical changes and treatment, 251-4

Jacks, 205
Jetsam, 42

Knots, 50

Lagan, 42
Landmarks, 96, 140
Law, 41, 42, 281-9
Layering, 159, 161
Leases, 45
Levelling, 87, 89, 171, 172
Life Saving Apparatus Volunteer Force, 40
Lifting, by boat, 230; by oil barrel, 231; by bags, 232
Line sinkers, 135
Lloyds, Corporation of, 39; Lists, 39; Register of Shipping, 39; Underwriters, 38, 44

Magnetometer, 121-3
Mahdia, 161, 210
Markers, bottom, 50, 62, 64, 112, 164, 176; floating, 57, 59, 61, 103, 114, 176
Mary Rose, 25-6, 126
Measuring Chain, 75
Merchant Shipping Act, (1846), 42; (1894), 43-4; (1906), 43
Metal detectors, 120-2
Metric scale, 52
Ministry of Defence, 15
Morrison Grid, 68-72, 186-7

Index

Moulds, 132, 251-3
Multiple fix, 142
Museum, Truro, 40

Nora, Sardinia, 31, 33, 138, 180

Paints, 56, 90
Parliament, (Act of 1841), 41
Parliamentary Committee Reports, 39
Pencils, 53
Perspective, 71, 190-2
Perspex, 91
Pewter, 256
Photographs, aerial, 47-8, 136, 180; balloon, 183; calculating area of, 188; focusing, 201
Photographic coverage, 64, 71, 159, 178-9, 183-4, 198-9; mosaic, 187
Photogrammetry, 69, 72, 189-90, 191-3; stereo, 189
Picket posts, 61
Pinger, see sub-bottom sonar
Pitons, iron, 61
Plane tables, 51, 78, 81, 89, 167-8; levelling by, 168-9
Plumb lines, 73, 184-6
Polystyrene, 56
Polythene, 90
Poly Vinyl Chloride (PVC), 52-3, 57, 69, 80
Portable square, 193; correction for slope, 195
Port Books, 39
Port Royal, Jamaica, 16
Pottery, identification from, 245
Preservation order, 46
Probes, 115
Protractor, 82, 85, 87, 147, 149, 157
Publication, 263-7
Pythagoras, Theorem of, 67, 75

Quays, 32

Ranging poles and staffs, 73, 89, 154, 168

Receiver of Wreck, see wreck
Recording, 177
Reservoir for aqua level, 91-2
Right angles, laying, 75
Roadsteads, see anchorages
Royal George, 26

Sacks, 59
Salvage, laws of, 14, 41-5; agreements, 45; first salvor, 43-4; injunction against, 44-5
Santa Maria de la Rosa, 15, 35, 41, 44, 100, 130
Scaffolding poles, 67
School for Nautical Archaeology (SNAP), 16, 40, 131
Sea level changes, causes of, 31-2
Seals, amphora, 27
Search, swim line, 96-103; spacing, 99; square, 111; circular, 111; sweep, 113-14
Sextant, 51, 78, 143-9
Sonar, 123; side scan, 125-6; sub-bottom, 126-7
Sparker, 127
Stadia lines, 153-4
Sunburn, 98
Surveying, 149; stations, 151, 155

Tacheometry, 153-4
Tapes, measuring, 50, 75, 77, 85, 164-7
Teredo Navalis, see shipworm
Theodolite, 51, 59, 78, 82, 149, 152, 153-4; underwater, 82-7, 171; zeroing, 156, 171
Tides, effect of, 46, 62, 88, 95, 103
Timbers, ships, 133, 242
Times, The, 39
Tobacco pipes, 246
Tow lines, 109
Trade routes, 27
Transits, 47, 80, 96, 100, 102-3, 141; Photographic, 141
Treasure Trove, 45
Triangulation, also trilateration, 50, 53, 114, 162-5

Trinity House, 39
Trunnels, 243

Underwater Association, 72

Vasa (Wasa), 19, 23, 25, 134, 230
Vertical staff, 154
Victory, HMS, 19
Visibility, 96, 99, 136

Vision, arc of, 96

Wasa, see Vasa
Water jet, 217-20; pumps for, 219
Water dredge, 221-4; capacities, 224
Weights, scale, 245
Wreck, 39, 42-5, 118, 130, 135, 160; Receiver of, 39, 43-5